P9-DVB-873

HEALTH CARE PROVIDERS, INSTITUTIONS, AND PATIENTS: CHANGING PATTERNS OF CARE PROVISION AND CARE DELIVERY

Research in the Sociology of Health Care
Jennie Jacobs Kronenfeld, Series Editor

(*Volumes 10 and 11 edited by Jennie Jacobs Kronenfeld and Rose Weitz*)

Volume 10: 1993

Volume 11: *Agents of Health and Illness*, 1994

(*Volumes 12–16 edited by Jennie Jacobs Kronenfeld*)

Volume 12: 1995

Volume 13: *Health Care Delivery System Changes: New Roles for Providers, Insurers, and Patients*, 1996

Volume 14: *The Evolving Health Care Delivery System: Necessary Changes for Providers of Care, Consumers and Patients*, 1997

Volume 15: *Changing Organizational Forms of Delivering Health Care: The Impact of Managed Care and Other Changes on Patients and Providers*, 1998

Volume 16: *Quality, Planning of Services, and Access Concerns: Impacts on Providers of Care, Health Care Institutions, and Patients*, 1999

HEALTH CARE PROVIDERS, INSTITUTIONS, AND PATIENTS: CHANGING PATTERNS OF CARE PROVISION AND CARE DELIVERY

Edited by   JENNIE JACOBS KRONENFELD
Department of Sociology
Arizona State University

JAI PRESS INC.
Stamford, Connecticut

ISBN: 0-7623-0644-0
ISSN: 0275-4959

Manufactured in the United States of America

CONTENTS

PART III.   GENDER AND HEALTH

PART IV.   RESTRUCTURING OF CARE

LIST OF CONTRIBUTORS

Gary L. Albrecht

School of Public Health
University of Illinois
Chicago, Illinois

Letitia T. Alston

George Bush School of Government
  and Public Service
Center for Public Leadership Studies
Texas A&M University
College Station, Texas

Cornelia Beck

Departments of Geriatrics and
  of Psychiatry and
  Behavioral Sciences
College of Medicine
University of Arkansas for
  Medical Sciences
Little Rock, Arkansas

John R. Booher

The Associates
Irving, Texas

Brenda M. Booth

Centers for Mental Health and
  Outcomes Research
Department of Psychiatry
University of Arkansas for Medical
  Sciences
Little Rock, Arkansas

Jin-Yuan Chern

Department of Health
  Administration
Chang Jung University
Tainan, Taiwan

Neale R. Chumbler   Centers for Health Services Research
                                     Marshfield Medical Research
                                         Foundation
                                     Marshfield, Wisconsin

Marisue Cody   Centers for Mental Health and
                                         Outcomes Research
                                     Central Arkansas Veterans
                                         Healthcare System
                                     Little Rock, Arkansas

Jeff Erger   Department of Sociology
                                     University of California
                                     Los Angeles, California

Elizabeth Furlong   School of Nursing
                                     Creighton University
                                     Omaha, Nebraska

Melinda Goldner   Sociology Department
                                     Union College
                                     Schenectady, New York

Oscar Grusky   Department of Sociology
                                     University of California
                                     Los Angeles, California

Shirley Harkess   Department of Sociology
                                     University of Kansas
                                     Lawrence, Kansas

Michael I. Harrison   Department of Sociology and
                                         Anthropology
                                     Bar-Ilan University
                                     Ramat-Gan, Israel

Kristen Karlberg   Department of Social and Behavioral
                                         Sciences
                                     University of California
                                     San Francisco, California

Jennie Jacobs Kronenfeld

Department of Sociology
Arizona State University
Tempe, Arizona

Harm Lieverdink

Supervisory Board for Health Care
  Insurance (CTU)
Amstelveen, Netherlands

William Alex McIntosh

Center for Public Leadership Studies
George Bush School of Government
  and Public Service
Texas A&M University
College Station, Texas

Traci Mann

Department of Psychology
University of California
Los Angeles, California

William D. Marelich

Department of Psychology
California State University
Fullerton, California

Kathleen Johnston Roberts

Department of Sociology
University of California
Los Angeles, California

Jerry Ross

Department of World Business
The American Graduate School of
  International Management
Glendale, Arizona

Louis F. Rossiter

Department of Health
  Administration
Virginia Commonwealth University
Richmond, Virginia

Clasina B. Segura

Department of Sociology
Texas A&M University
College Station, Texas

Janet K. Shim

Department of Social and Behavioral
  Sciences
University of California,
  San Francisco

Dianne Sykes

Department of Sociology
Texas A&M University
College Station, Texas

Jan E. Thomas

Department of Sociology
Georgetown University
Washington, D.C.

Thomas T.H. Wan

Department of Health
  Administration
Virginia Commonwealth University
Richmond, Virginia

Marlene Wilken

School of Nursing
Creighton University
Omaha, Nebraska

Mary K. Zimmerman

Department of Health Policy and
  Management
University of Kansas
Lawrence, Kansas

INTRODUCTION

Jennie Jacobs Kronenfeld

This volume explores issues connected with health care providers, institutions, and patients. The focus of many of the articles is on changing patterns of care delivery and provision of care, as it affects these important groups of actors within the health care system. The articles range from those that focus on more specialized groups of patients such as the elderly to those that focus on people who deliver health care services to those that deal with more general issues of the restructuring of the U.S. health care system.

This volume is the first one in this series in the new century. The last decade of the twentieth century has been one of much ferment and change within the health care system. The decade began with another proposed major overhaul of the U.S. health care system, the Clinton health care reform plan, which would have led to major changes in the organization of health care and the structure of health care delivery, with a central role for the federal government. This plan failed as have all previous efforts at major overhaul of the U.S. health care delivery system, although various types of piecemeal reforms and expansion of access to health insurance have occurred in the U.S., especially Medicare for the elderly and

Medicaid for the categorically poor in 1965. Other expansions have been coverage for those with end-stage renal failure and those who are classified by Social Security as disabled. Most recently, a new title has been added to the Medicare and Medicaid legislation, the CHIP (Child Health Insurance Program) that provides health insurance and access to care to many children of the working poor. Despite these modest expansions, the United States remains the only industrialized country in the world that does not have some type of guarantee of coverage of health care to its citizens.

By the end of the 90s, many of the types of changes proposed by the Clinton plan were occurring, but as part of marketplace-driven reform rather than government reform. The most important of these trends is the current growth in managed care in the United States along with impacts that has had on the role of providers in the health care system, not only physicians but also nurses and other less traditional types such as genetic care providers. Some of the issues relating to these providers are addressed by papers in this volume. Market driven reforms have occurred in countries outside the United States also, as one of the papers in this volume indicates in its discussion of The Netherlands and its health care delivery system.

Market driven reforms impact not only providers of care, but also patients. Not all patients experience the same kinds of problems or concerns under market reforms. Some groups may be more vulnerable, such as the elderly, the terminally ill, or those with special challenges such as mental impairments. Women may experience special challenges in this changing health care system. The first part of this volume deals with the elderly and terminally ill. Several of the papers herein explore issues linked to gender such as gender bias in health care, the impact of feminist health centers, and the overall impact of feminism on medical sociology as a field and on biomedical knowledge more broadly.

Many health policy experts as well as medical sociologists now believe that the major changes to the US health care system in the last decade of the twentieth century have been due to the growth in managed care, often as part of a response to concerns about costs of health care services. These changes have been pushed by employers and by insurance companies, both of which are looking for ways to have reasonable quality health care delivered at reasonable costs. Sometimes these changes have been pushed by states (as in the introduction of more managed care in the Medicaid program), often with the goal of holding down costs in that particular state. For states, the goal is often to limit the amount of resources that the state has to devote to the state portion of the Medicaid match (matched by federal dollars), the major way in which the Medicaid program operates. Some major changes are being pushed by the federal government as well as consumer policy groups, such as the passage of the CHIP, which will allow states to provide health care insurance to a broader range of children in the state if the specific state so chooses. Within the United States, the most pronounced conflicts within the health care system have been between private providers and government policy makers and their constituents. A different way to think about this conflict is as one between profit versus coverage. Many of the policy oriented analyses of the last decade have focused on these issues. While these are important, the next decade may see other issues in changing patterns of care provision as just as important. A trend in that direction in this volume is the attention provided to issues of gender, of aging populations, and of new approaches to delivery of care such as telemedicine or integrative medicine.

Cost, quality, and access have been the three major watchwords of the health care system in the United States for the past 30 years. Of these three major themes, sociologists tended to focus more heavily on issues of quality and access rather than on cost, often leaving this to economists. Most of the papers continue in these directions, with a greater focus on issues of quality and access to care. Quality is often linked to special groups of patients, and issues of quality and access are linked to gender concerns and to concerns about different models of care delivery than those that were the dominant ones for most of the second decade of the twentieth century.

This volume focuses upon scholarly work in the sociology of health care. Unlike quick policy analysis studies conducted by think tanks, evaluations undertaken for health maintenance organizations (HMOs) or state health departments, or reviews and analyses of problems of health care providers written by researchers from within those fields (such as physicians or nurses), scholarly work in medical sociology comes from years of thought, data collection efforts, and data analysis. The work often follows the tradition within sociology of looking at things with a critical eye, with an unwillingness to accept what is already there as the best or most appropriate. Thus medical sociological work often challenges accepted conventions, beliefs, and approaches. This book continues in that proud tradition of work in medical sociology. Given the time frame, this work often begins long before political changes or overall system changes occur. Thus all of the work in this volume was begun before the current debate over the passage of a patients' bill of rights to help patients deal with the power of managed care organizations and HMOs over what care is actually provided to patients. Good sociological research, however, even though it may have begun before the current "hot" health care issues, will often link to current health care issues. The research will tie in to a tradition of research in medical sociology and to theoretical concerns that the discipline has addressed for a number of years. This is true of many of these chapters, which challenge the current health care system, whether as questions about how care is provided to the terminally ill or how care is provided to those with memory-related problems, or by raising questions about gender bias in medical education and linkages of race, class, and gender to biomedical knowledge.

The chapters in Volume 17 are divided into four parts. The first deals with care for the elderly and terminally ill. These are very important concerns for the health care system of the future, as one of the major challenges for all health care systems in industrialized countries, including the United States, is to figure out how to provide high quality, accessible, reasonably priced services to the growing elderly and disabled populations of these countries. This part contains only two chapters, but both relate to these challenges and the implications for health care providers, institutions, and the patients themselves.

Ross and Albrecht focus on the care of the terminally ill. They begin with an appreciation of the importance of excessive medical interventions that have become accepted patterns of care provision for the terminally ill. They point out that common explanations for these issues in the past have focused upon the impact of physician decisions and the shift over the last 50 years so that financial costs are borne more and more by third parties and less and less by the patient and his or her family. This paper presents a more detailed and carefully constructed view of this issue, with greater focus on how these situations often begin with a rational calculus of what is appropriate. At some point, ranges of personal, psychological, social, and organizational factors all interact to escalate commitment to a failing course of action. One possible solution would be a new specialty with a greater focus upon palliative care and ethics, along with requisite technical expertise.

Care of the terminally ill is one of the major challenges of an aging population. The other major challenge is dealing with those with declining memories as they age, which is the focus of the chapter by Chumbler and colleagues. It makes an investigation into the interrelationships among levels of education—residence, physical and mental status, and the use of primary care physicians for memory problems. Patients use of care patterns vary by factors such as where they live. Rural patients have more memory impairments but similar rates of use of primary care physicians for the memory problems. Older adults with memory problems are underserved in many communities, perhaps moreso for those in rural areas. Primary care physicians in the new century may need more training to recognize the seriousness of memory impairment and more information about specialized services available to patients with these problems. While some researchers are upbeat about finding genetic solutions to memory loss, until such time as this occurs, improved recognition of these problems and improved systems for treating them will be necessary.

The four chapters within the second part of this volume focus on health professions and occupations. Health professions and occupations are very important in the current changes in health care delivery. Many researchers argue that the growth of managed care is leading to less autonomy for physicians and is lessening their satisfaction with their careers. The relative roles of generalists and specialists are also changing. Nurses are also undertaking new tasks and finding new challenges in changes in the health care delivery system. Both managed care and changes in diseases and in care patterns lead to changes in health professions. New groups emerge such as genetic care providers and specialists who deal with AIDS. All of these topics are addressed in this section.

Furlong and Wilken describe how the autonomy of nursing is being threatened by changes in the health care delivery system. Many papers in the literature have discussed the issue of loss of autonomy for physicians. This is one of the first in medical sociology to discuss a related issue for nursing. Licensing and credentialing of health professions with an employing organization rather than through a traditional state profession-specific regulatory board is the major issue discussed. They also contrast how trends within one health profession (in their case, nursing) may increase power and autonomy of certain subgroups but threaten the overall profession.

Changes in health care professions are not happening only in the United States. In the second paper in this section, Harrison and Lieverdink explore the current situation for physicians in The Netherlands. They focus on one particular time period (1982 through 1992) and one dramatic instance in which physicians resisted managerial and governmental controls. Hospital specialists were able to block efforts by both government regulators and hospital managers to control the activities of the specialists. The chapter includes a discussion of four types of professional controls and how recent developments in The Netherlands have affected these.

Not all changes for health providers are linked with managed care and attempts to control costs. Some link to changes in technology and disease patterns. This is true for the last two papers in this section. Karlberg's work on genetics care providers presents both participant observation and semi-structured interview data that helps the reader understand how this new field has developed and the tensions that result from the uncertainties and ambiguities in medical genetics. Marelich and his colleagues present a two-stage model of decision making that is used by health care providers dealing with AIDS patients, focusing upon the decision to use combined antiretroviral therapy. The paper forms the model after interviewing both AIDS patients and providers working within this area. The more that decisions included behavioral and nonmedical components, the more they presented difficulty to the providers. Because the providers with the most contact with the patients who were interviewed in this study were nurse practitioners and physicians' assistants, the high level of decision making combined with relatively low authority to create decision making uncertainty.

Part III includes four chapters dealing with various aspects of gender and health. In sociology in the last two decades, research on gender has changed the field and broadened understanding of society. The incorporation of the gender lens to deepen our understanding within medical sociology is also important, and the chapters of this part help to do it in various ways. Mary Zimmerman's paper concentrates on women's health and gender bias in medical education. She explores the connections between the Women's Health Movement as the first significant and broad-based critique of modern medicine and the presence of women today within medicine, as medical students, faculty, and players in the decision-making structures of U.S. medical schools. This includes an examination of gender bias in medical curricula. More curriculum change should occur in the future as the result of some major research that includes women as patients is completed.

Thomas' paper focuses on a particular type of health care delivery, the feminist women's health centers. These centers were part of that initial thrust of the Women's Health Movement. These centers have formed new models of patient-centered care and patient empowerment. Fourteen different feminist women's health centers have been the sites of the studies composing the research described in this chapter. A special focus of interest is the strategies used by the different centers to empower patients.

The third chapter in this section is more inward looking, focusing upon the impact of feminism on mainstream medical sociology. Some fields within sociology, such as family and work, have been transformed fundamentally by feminism and by an incorporation of gender into their subfields. Harkess examines representative research in mainstream medical sociology and determines to what extent it has been influenced by feminism, with a focus on research published in the *Journal of Health and Social Behavior* along with a few articles in more general sociology journals that focused on medical and health related topics. She finds mainstream medical sociology less transformed than some other subfields.

Shim's paper tackles a very broad subject by exploring bio-power and racial, class, and gender formation as linked to biomedical knowledge production. She reviews a very vibrant debate not only within medical sociology but also within some fields more closely connected with medical research such as epidemiology. Within that field, questioning of categories such as class, gender, and race has been occurring. Shim brings a more sociological perspective to this discussion and especially emphasizes the importance of remembering that risks are embedded in and shaped by structural conditions. Race and gender describe social relations of power, not simply attributes linked to the body of a person. The relational processes are essential to a more sociological understanding of these issues.

Part IV contains three chapters, each dealing with a specific aspect of restructuring of care. The first, handled by McIntosh and his colleagues, deals with predictors of use of telemedicine for differing medical conditions. The chapter uses a framework based on health care utilization and adoption of innovation models to compare patients who have experienced telemedicine. Goldner then deals with integrative medicine: the integration of alternative practitioners with Western medical techniques. She links some of the trends in growth of alternative medicine and especially of growing interest in alternative medicine among physicians with the growth of managed care and its attendant erosion of physician authority. She believes that some physicians are searching for new ways to practice medicine without those structural constraints while others are attracted by the connection that alternative ideology makes between spirituality and medical practice. The final chapter uses quantitative analytical techniques with a three-stage, cross-lagged model incorporating a longitudinal design to determine the impact of prior utilization on subsequent utilization. Both a direct and indirect effect occurs, with prior health status playing a role.

All of the contributions to this volume relate to health providers, institutions that provide health care, and the care that patients receive from these providers and institutions. They are a diverse literature in some ways, ranging from specific concerns of the elderly to concerns about gender and health to topics relating to providers of care to topics dealing with restructuring of care. One interesting point is that the book's final section dealing with restructuring of care approaches the subject not in an economic sense, which is often the angle of more popular discussions in the general media, but in terms of issues such as telemedicine, integrative medicine, and utilization of care. These chapters reflect not only the current diversity in the sociology of health care but also the tendency of medical sociologists to critically look beyond the immediate focus and into the broader context of issues. They offer innovative and creative examination of important matters in the sociology of health care today and for the new century.

PART I

CARE FOR THE ELDERLY AND TERMINALLY ILL

UNDERSTANDING AND MANAGING HEALTH CARE INTERVENTIONS IN THE TERMINALLY ILL

Jerry Ross and Gary L. Albrecht

ABSTRACT

There is a growing consensus that there are often excessive medical interventions in terminally ill patients. This problem is usually seen as stemming from physician decisions in applying new technology in a context in which financial costs have been borne by third parties. We believe this is, at best, a partial explanation for the phenomenon. The tendency to escalate commitment—to persist in failing courses of action—has been found by social scientists to occur in a wide variety of decision contexts. In ethnographically examining health care interventions in terminally ill patients, we found that a wide range of rational calculus, psychological, social, organizational, and contextual factors interact over time to contribute to excessive persistence. Intervention decisions reflect a complex, fluid interplay between patients, health care providers, institutions, and an array of external stakeholders. Effective revisions of current patterns of care practices must address the nature and complexity of the sources of the problem. We suggest a series of strategies including a new medical specialty to deal with these issues.

Research in the Sociology of Health Care, Volume 17, pages 3-29.
Copyright © 2000 by JAI Press Inc.
All rights of reproduction in any form reserved.
ISBN: 0-7623-0644-0

INTRODUCTION

A major social debate is underway over the appropriate level of medical intervention in terminally ill patients. There is substantial unhappiness both in the general public and among many health care experts over how these intervention decisions are now made. It is widely believed that many medical interventions are excessive and need to be curtailed (Miller, Quill, Brody, Fletcher, Gostin, and Miller 1994; Nuland 1994; Schneiderman and Jecker 1995).

Some of this dissatisfaction reflects unease with the application of the ever-increasing technologies available to health care professionals. There is growing evidence that many intensive interventions during the last years of life such as the use of hemodynamic therapy do not seem of any value in reducing mortality or morbidity (Gattinoni et al. 1995). A variety of commonly performed procedures such as right heart catherization have also been shown to be of questionable efficacy. David Eddy (1993) estimated that 80% to 90% of all medical treatments have not been adequately evaluated in controlled studies.

Another source of dissatisfaction stems from an apparent lack of cost–benefit assessments of these interventions. Numerous studies point out the high level of expenditures that occur during the last year of patients' lives (Scitovsky 1994). Revisionists' perspectives on this issue have emphasized that this dilemma has a limited impact on the nation's overall health care bill. Yet even these critics place the annual price tag for such costs at $60 billion a year (Emanuel and Emanuel 1994). Additional "costs" include the pain, discomfort, and intrusiveness of many procedures, procedures whose benefits in prolonging life are often highly questionable.

Perhaps the greatest frustration, however, arises from current health care practices' failure to pay adequate attention to the wishes of the patient and the patient's family. For example, a large clinical trial at five teaching hospitals in the United States designed to examine end-of-life decision making documented shortcomings in communication, treatments, and circumstances of hospital death (Support Principal Investigators 1995). The research discovered that only 47% of physicians knew when their patients preferred to avoid cardiopulmonary resuscitation; 46% of "Do Not Resuscitate" codes were written two days or less before death; 38% of the patients who died spent at least 10 days in an intensive care unit; and for 50% of the conscious patients who died in the hospital, family members reported moderate to severe pain at least half of the time.

The debate over controlling excessive medical interventions is unlikely to be resolved soon. The U.S. population is aging and the technology of medical interventions continues to expand. Some observers suggest that better evaluations of the efficacy of new medical technology will substantially curb excessive interventions. Others argue that with the movement to HMOs, the changing financial incentives against intervention will cure the bias toward intervention. Still other observers point to legal victories in the right-to-die movement as evidence that excessive intervention may soon be a thing of the past. Yet we believe that many of the current attempts to intervene in the process and bring it under control are destined to be failures. Generally those examining the intervention process focus almost exclusively on the doctor–patient dyad and do so without explicating a theory for why persistent interventions are so common. Thus, there is an attempt to solve a problem without first understanding what causes it.

The tendency to invest more resources in failing courses of action is by no means unique to the medical arena. Over the last 20 years, a substantial body of social science research has documented a general tendency for individuals and organizations to invest more resources—to escalate commitment—in response to unsuccessful courses of action (for a review, see Staw and Ross 1989).

We believe that such a perspective can usefully be applied to gain a more complete understanding of how medical intervention and treatment decisions are made in the terminally ill. Using the variables and categories suggested by previous escalation research, we provide illustrations based on our previous experiences and observations; from interviews and discussions with patients, health care providers, family members, and other relevant parties; and from prior health care literature. Our aim is to provide a fuller and more accurate explanation for excessive interventions, to augment our ability to predict these interventions, and ultimately to provide assistance in controlling excessive interventions.

ESCALATION THEORY

Escalation theory suggests that there are systematic factors that lead individuals and organizations to persist with failing courses of action. "Escalation theory" was originally developed in the management area to explain excessive persistence with financial investments. Staw's (1976) inaugural study examined the role of personal responsibility in leading financial decision-makers to increase their monetary investments to a selected course of action. Staw found that the decision makers who had high levels of personal responsibility for failing endeavors were more likely to respond by increasing their investments than were those with low levels of personal responsibility. Staw attributed these results to self-justification: the decision makers were unwilling to admit to themselves that their initial decision had been a mistake.

During the next two decades, a substantial body of research arose identifying determinants of excessive persistence. Reviews of this literature typically group the determinants into five classes: project, psychological, social, organizational, and contextual. Project variables pertain to the objective features of the project. In a for-profit realm, these would include primarily economic factors that would affect the rational calculus of abandoning versus redoubling one's effort. Examples would be whether the setback was viewed as temporary versus permanent, the closing costs and salvage value of prior investments, and the economic merits (costs vs. benefits) of pursuing a particular course of action (for examples of research on project variables, see Bateman 1983; McCain 1986; Northcraft and Wolf 1984).

Psychological sources of persistence include reinforcement traps, which create difficulties in discontinuing previously rewarded activities, and individual motivations such as the need for self-justification. Decision making errors such as sunk costs effects and biases in information processing such as the tendency to slant data in the direction of preexisting beliefs can also lead individuals to persist in a course of action (for examples of research on psychological determinants, see Arkes and Blumer 1985; Bobocel and Meyer 1994; Conlon and Parks 1987; Goltz 1993; Staw 1976).

Social determinants are interpersonal processes that can lead to excessive commitment such as desire to justify losing projects to potentially hostile audiences, modeling of others' behavior in similar circumstances, and norms for consistent or strong leadership. For examples, see Brockner and Rubin (1985), Fox and Staw (1979), and Staw and Ross (1980).

Organizational sources of commitment include variables such as the level of political support for a project within the organization and whether economic or technical side bets such as the hiring of staff or development of unique expertise have taken place. Courses of action may also become institutionalized as they are tied to the organizations' central values and objectives. Extraorganizational parties may also become involved as commitment episodes unfold over time. Particularly in late-stage events, the relative balance of constituencies favoring persistence and withdrawal may become critical for determining escalation (for examples, see Goodman, Bazerman, and Conlon 1980; March 1978; Pfeffer 1981; Ross and Albrecht 1995; Ross and Staw 1993).

Thus, we see from this very brief review that over the last two decades the study of escalation of interventions has involved a considerable broadening of focus. From early attempts to understand excessive persistence as stemming from a single source at a single time period, researchers have been forced to recognize that escalation of commitment is often a longitudinal, multidetermined phenomenon caused by a range of factors associated with the utility of the intervention itself as well as psychological, social, organizational, and external forces. Researchers have found that only by considering the full range of participants involved and by examining the broad context of ongoing decisions can a full understanding of why persistence with failing courses of action takes place be obtained

METHODS

This paper is the result of a cross-disciplinary collaboration between a management theorist with over twenty years of experimental and field research experience in decision making and a medical sociologist with more than 30 years of experience in health care delivery. Starting with our disparate theoretical and research backgrounds, we gathered field data via ethnography, indepth interviews, and focus groups with physicians and patients in emergency departments, departments of general internal medicine, and private practice clinics in the United States over a two-year period.

Ethnographies were conducted in three different emergency departments and three departments of general internal medicine located in a large urban public hospital, a community teaching hospital, and a university teaching hospital. A major part of this work was studying how medical care providers, the patient, and family members dealt with life threatening and terminal illnesses. During the research period, the ethnographer observed the daily routine of the clinic, the organization and flow of the work, the interactions of the staff, patient-physician encounters, and the ways that patients in terminal conditions were diagnosed, treated, and referred to other professionals or facilities. Special attention was given to seeing how the family or patient was involved in the communication and decision processes. After the ethnographer became familiar with the organizational routines, indepth interviews were conducted with health care staff, family members, and the patient about the course of treatment, the rationale behind the treatment, and the patient and family feelings about the entire process.

Two focus groups were later conducted with patients with serious chronic illnesses and their families to explore their experience with confronting a life-threatening illness and potential death.

In addition, one of the authors interviewed 137 persons with serious disabilities and some of their family members about managing chronic illness and acute episodes during a terminal illness. Attention was given in the semistructured interviews to patients' and family's accounts of their participation in decisions about medical interventions and the terminal patient's care, living situation, and general quality of life.

Also, during this time period, one of the authors was a major support to his father and another relative during their dying process. While spending hours with the dying patients and relatives in hospitals, clinics, and nursing homes, he kept field notes on his observations about the way that their terminal illnesses were defined and managed and the way the patients and his relatives responded to the actions and decisions of health care professionals. The experience of daily living with the deteriorating health and quality of life of one's own relatives provided an emotional and personal context to the dying process.

Based upon these data sources and prior theory, we provide a broad explication of factors influencing treatment interventions in terminally ill patients. We examine not only the rational calculus of intervention, but also psychological, social, organization, and contextual factors. We begin with a discussion of the rational calculus of intervention.

THE RATIONAL CALCULUS OF INTERVENTION

The narrow rational calculus of a medical intervention most commonly focuses on a joint decision by the physician and the patient. In theory, the predominant voice is that of the patient and is heavily influenced by the patient's value system (Kuntz 1998). Some patients clearly believe that there is nothing worse than death and wish to use every possible means to avoid it, while others regard dying as a natural step in the life process. So, for example, even after enduring four major resections of her colon for cancer which left her in continual pain and without control of her bowels, an eighty-two year old wealthy woman insisted to her family and doctor that, "Even if I lose consciousness, I want you to take every medical step to keep me alive regardless of the cost." Her tertiary care physicians said to her husband, "You know we can only keep her alive for a few extra months and it will not be a pleasant experience for her." The husband acknowledged their judgement but asked them to persist because of her wishes.

On the other hand, a 48-year-old male with metastatic cancer of the liver responded when confronted with the lab tests and the doctors' prognosis of terminal disease, "My mother taught me when I was young that when your time was up, God would call you. I guess my time is up, so please let me say goodbye and pass."

Obviously the nature of the patient's ailments influences calculations regarding intervention. Diseases like pancreatic and liver cancers are almost uniformly fatal, often in the short run. Other diseases may take a long time and may also be much more likely to be cured. In more ambiguous instances a number of factors associated with the patient's illness influence the decision to commence aggressive interventions versus offering only palliative assistance.

The "cost" of the treatment certainly factors into the decision. In some instances the "costs" are not financial but have to do with the amount of energy required to move a patient back and forth between formal and informal health care settings, pain, discomfort, and loss of dignity to the patient, and the caretaker burden (Haug 1990). As in other realms, "investments" with potential longterm payoffs are likely to induce substantial investment—thus extraordinary efforts extended in addressing childhood leukemia or countering difficulties in premature babies. These are in contrast with the situations of older patients needing heart or kidney transplants. Different interventions may also have different structures of salvage value and closing costs. For many kinds of cancer treatments there are reasonable salvage values. Even if the first rounds of chemotherapy do not produce long-term remissions, the patient may gain some additional time. Other interventions may either produce a cure or perhaps leave the patient worse off than before.

The physician's value system may generally heavily influence treatment decisions. Patients look to physicians to frame results of the tests and provide guidance about how to proceed. Yet how many of us really know our doctors well? How many of us have had even a 10-minute conversation with our physicians

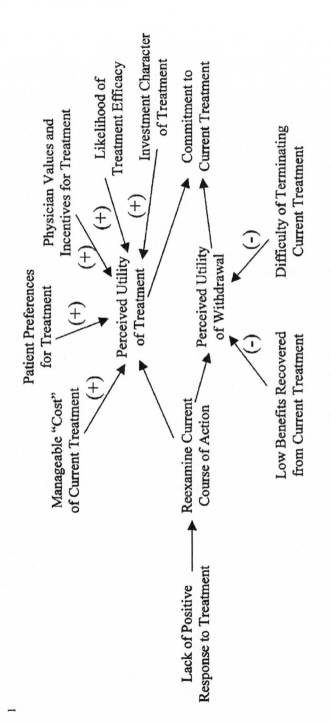

Figure 1. The "Rational Calculus" of continued Intervention

9

about their philosophies of intervention? Further, as Poplin (1996) points out, managed care is not only changing the financial incentives toward withdrawal of treatment versus continuance of treatment, it is also changing the level of personal bonds between doctors and their patients. The result may be even less accurate communication between patients and their physicians.

Yet often it is the physician's decision that determines the course of treatment. Quill and Bennett (1992) found that less than 25% of Cardiopulmonary Resuscitation (CPR) and "Do Not Resuscitate" (DNR) decisions made in a New York hospital were the result of informed decision making by the patients themselves. Instead someone else has to decide whether to intervene or alternatively to cease treatment. Physicians in these positions often delay the decision or defer it to someone else. A group of family members in the critical care waiting room said to the attending physician, "Stephen has suffered enough. We trust you to do what's right." This places a terrible burden on the physician (Orentlicher 1996). This burden is exacerbated in the Emergency Department situation where a patient arrives in critical condition without reliable medical information, previous medical history, or a family member or friend. The physician often does not know what the wishes of the patient or family are and has to act based on the sketchy information available and dynamics of the situation. A physician responding to these pressures said, "We try to stabilize the patient and contact the relatives but we can't wait. Some of these people are homeless or without an address. We have to play physician, family, and guess what the patient would want if they knew what was going on. This makes all of us uncomfortable regardless of our experience."

A common solution to dealing with uncertainty about the philosophy of the patient or the patient's relatives is for the physician to adopt a clinical model in which the treatment of the ailments is a technical problem. The patient's wishes or overall situation can easily become disregarded when this perspective is adopted.

An emerging area of debate has to do with the physician's obligations when the patient or the patient's family insists on treatment that is known to be medically futile. Schneiderman and Jecker (1995, pp. 155-156) report on the Baby K case, in which a baby born with most of her brain, skull, and scalp missing, nevertheless received all available aggressive life-support treatments. One of the difficulties in resolving these cases is evidence that physicians vary widely in their operational definition of futility (Swanson and McCrary 1996).

Ideally physicians are to act as agents working in the best interest of the patients. Yet, as in other agency contexts, the physician's and patient's best interests do not always completely coincide. Some of the issues are financial. Historically most medical interventions have had a fee-for-service financial incentive. As one might expect, such incentives tilt the scales for treatment, although the amount of this incentive effect is unclear. Sharpe and Faden (1996) believe that, as concerns about cost containment provide incentives to cut back on expensive services, the problem of underuse may surpass overuse or unnecessary treatment as a serious risk to patients.

Physicians may also need to balance the current patient's interest with the interests of other patients and the general state of medical knowledge. In one instance we observed, a researcher/clinician at a major national medical school said to his resident on reviewing the declining progress of a terminal AIDS patient, "Let's add a third course of drugs because he meets the case eligible criteria for our research protocol."

Thus, the decision to initially intervene is a complex one involving an interplay between the patient's values and preferences, the nature of the illness, and the physician's value system and incentives. When an intervention is not initially successful, the decision-makers must decide whether to cease treatment or try still additional measures. As in other situations, the decision to persist in medical interventions is attributable to the interplay of not only the narrow rational calculus of intervention but also with various psychological, social, organizational, and contextual factors operating on the decision makers.

PSYCHOLOGICAL FACTORS
INFLUENCING PERSISTENCE

Two psychological variables that appear fundamental are basic reinforcement effects and biased information processing. Reinforcement effects operate on physicians by both producing a general tendency to intervene and fostering a tendency to persist with prior interventions. As they are trained, physicians are rewarded not merely for learning knowledge and procedures, but for applying them. It is thus natural for them to tend toward intervention when they encounter new patients. In one interview, a young resident described his use of all his recent class and residency training to revive a patient who had all the manifestations of being clinically dead. It was a stirring 20-minute monologue. At the end one of the researchers asked, almost as an afterthought, how the patient was doing now. The resident responded "Oh, he died the next night. He was clearly terminal, you know." Even in less time-pressured circumstances and with more experienced physicians, Nuland (1994) suggests such interventions are the rule rather than the exception: "At that point, and even now, more than a decade later, I had never had a single experience in which an oncology consultation did not result in a recommendation to treat" (p. 237).

There are numerous forces driving the physician to approach medicine as a process not only of diagnosis, but also of intervention. As Barbour (1995) points out, "If doctors do not do something, the question may be raised that perhaps they have not earned their fee" (p. 152). Moreover, "To treat the patient as a person required time, energy, and personal involvement. These are all things in short supply" (p. 153).

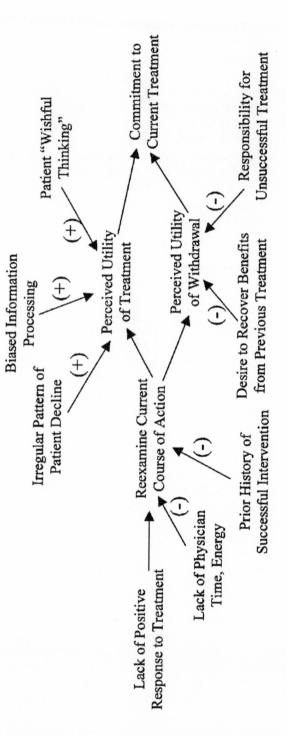

Figure 2. Psychological Sources of Continued Intervention

12

The structure of feedback about treatment interventions also contributes to persistence. Because reinforcement strengthens a behavior and intermittent scheduling makes behavior resistant to extinction, individuals may find it difficult to break out of behavior in erratic, slowly deteriorating courses of action. In many ailments, the patient's condition demonstrates not a steady continuous decline, but rather a downward trend with small occasional rallies or upswings. Both the patient and health care providers are susceptible to such variable interval reinforcement schedules. Charmaz (1991) characterizes these daily experiences as the "good days and bad days," which most commonly lead to a continuing decline in well being. Similarly, often the patient's level of suffering increases gradually. Those involved in the process—patient, family, and physician—become adapted to it. In one instance we observed, a terminal cancer patient experienced growing pain and discomfort that became increasingly resistant to palliative care. A relative arrived after not having seen the patient for two weeks and was appalled at the level of suffering. He furiously berated both the medical staff and the attending relatives for not insisting on and administering more pain killers. These daily observers had gradually grown accustomed to a reality that was a stark contrast to a more intermittent visitor.

Biased information processing may also contribute to escalation in a variety of ways. As summarized by Nisbett and Ross (1980) and Weiss and Bucuvalas (1980), individuals have an almost uncanny ability to bias facts in the direction of previously accepted beliefs and preferences. Physicians in many instances are psychologically resistant to admitting that the patient's condition is terminal. Often it is seen by physicians as an acknowledgement of their own personal failure since the patient has relied on the physician to keep themmalive (Fukkara, Tazewa, Nakajima, and Adochi 1995). The physicians may hesitate to admit to themselves that the patient is dying and they may likewise be hesitant to inform the patient.

The discussion of death is uncomfortable for most people including health care workers. Medical staff are generally idealistic about life and trained to preserve life through their knowledge, skills, and interventions. When a patient becomes clearly terminal to most close expert observers like the attending physician or floor nurse, the announcement and discussion of this judgement with other staff, the patient, and the patient's relatives is frequently avoided or strained. In the doctor's lounge of a tertiary care facility, a physician commented, "My job is to save lives. I can't support euthanasia. In fact, I can't even tell family members that their loved one isn't going to make it. I let somebody else do that."

Even when physicians bluntly convey the message, the patient may not hear it. In some instances, pain or distress may impede the communication. In many other instances, experienced physicians note the Freudian denial shown by patients when informed of the inevitably terminal nature of their conditions. There is, for instance, substantial evidence that cancer patients tend to overestimate their probability of longterm survival (Weeks et al. 1998). These misestimates also clearly influence the decisions that patients make about the kind of treatment they would like to receive.

Furthermore, even when patients accurately process prognosis data, the patient's response is not inevitably to curtail treatment. As in other domains, when people hear unwelcome information from one source they are likely to seek others. In some instances, this will involve simply seeking additional opinions within the medical establishment until a more favorable opinion is received. In other instances it will involve going outside the established medical practice to the vast gray area of herbalists, acupuncturists, psychics, laetrile clinics, and the like. Some of these alternative forms of care are palliative, but billions of dollars are spent on quackery every year, often by patients who would actively ridicule such activities prior to the onset of their own ailments.

In other instances, the patient is willing to acknowledge the futility of further interventions, but cannot get others to come to a similar acceptance. Patients' wishes are often not listened to by the doctor or family. When a patient says, "I want to say 'good-bye' and go" the statement is often dismissed as an aberrant request made by someone who is in pain, depressed, or confused. Escalation variables may shape the family member's and physician's behavior more than the patients' wishes or statements. A wife said about her critically ill husband who was declining further treatment, "Raymond has always been a fighter. If he really knew what he was doing, he'd want to fight through this right now." Yet while research indicates that patients' wishes to forego life-sustaining treatment are stable (Danis, Garrett, Harris, and Patrick 1994), it also shows that there is no evidence that patient preferences influence whether life-sustaining treatments are used (Danis et al. 1996; Schneiderman and Jecker 1995).

There are other psychological forces that may also operate both on patients and physicians and lead to persistence with medical interventions despite a lack of commensurate benefit. One such factor is self-justification: both physicians and patients may be unwilling to admit to themselves that their earlier intervention decisions may not have been wise ones. Thus, Ross and Albrecht (1995) found that doctors who initially selected aggressive interventions for pancreatic cancer were likely to continue courses of treatment despite lacking any evidence of positive clinical response.

A related contributor is what psychologists call sunk costs effects. Once major investments of time or money have been made in a course of action there is a natural tendency to want to recoup some benefit from the expenditure. In "rational" decision making, previous expenditures in a course of action should be irrelevant—they are sunk. The only factors that should influence our calculus are estimates about future expenditures and outcomes. Yet there is a massive literature that shows human beings typically do carry over previous expenditures in their mental calculus. Thus, having spent hundreds of thousands of dollars and endured substantial suffering undergoing an unsuccessful marrow transplant, an adult leukemia patient and her physician talked about still further interventions as a means of somehow making the marrow transplant pay off.

SOCIAL FACTORS INFLUENCING PERSISTENCE

In addition to these psychological sources of commitment, there are social forces that lead individuals to persist with failing interventions. These forces may produce continuance even if decision-makers have lost confidence in the possibility of a successful outcome. One such factor is external justification. In external justification, individuals may continue to invest in poorly performing courses of action not because they are unwilling to admit failure to themselves (as in instances of self-justification) but because they are unwilling to admit failure to other people.

Patients may feel the need to justify their decisions to others. In one instance we encountered, an elderly dying man asked his family for permission to die and begged that they not be angry at him for doing so. The man felt that accepting his own death would constitute failure to watch over and protect his wife, a role he had been playing for over 50 years. Several nurses told us it was not unusual for the terminally ill in the last stages not to die until their relatives went away, sometimes even just to the bathroom.

The norms of the sick role and health care community also demand that the patient make every attempt to live and additionally to be compliant (Wolinsky 1988). There are instances in which patients reported to us that when they refused to eat, several of the nurses became gruff with them, were physically less consoling, and were less quick to respond when other kinds of help were needed.

Physicians may also feel the need to justify actions to other physicians and health care professionals. Here the effect of local norms on health provider behavior may be substantial. Ross and Albrecht (1995) examined the physician decision making in four large urban hospitals in a case that involved a patient who was clearly in the terminal stages of pancreatic cancer. There was clearly variability in the treatment prescribed, variability that was most strongly associated with the site where the treatment was administered. In one hospital, residents making treatment decisions were far more likely to persist in a questionable course of action than those at the other locations. Interviews revealed that the socialization and control mechanism at that institution placed considerable pressure toward the continued delivery of services.

In these ambiguous situations of providing care to the terminally ill, one approach to decision making is modeling (Bandura 1977). Several studies have shown that individuals are likely both to select and to persist with a failing course of action when similar actions have been taken by others facing the same context. In medicine, this often occurs because students, interns, and residents model their behavior on attending physicians and are heavily influenced by the directions of the chief of medical staff and what is communicated during rounds. Furthermore, because of the threat of malpractice claims, physicians are influenced to take the more conservative treatment approach, which dictates intervention over possible withholding of treatment (Aaron 1994).

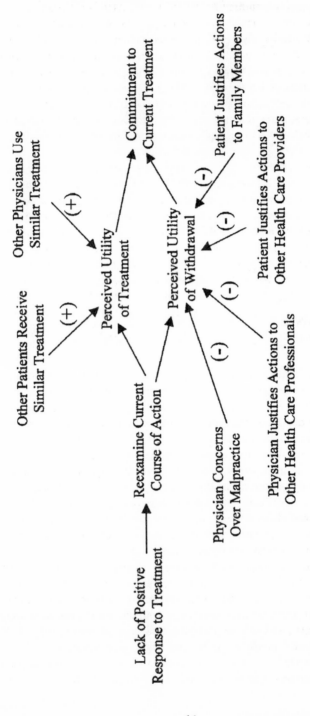

Figure 3. Social Sources of Continued Intervention

One of the factors that determines medical interventions is the constant threat of medical malpractice. Physicians constantly need to be asking themselves not merely what does the patient wish or what is in the patient's best interest but also what actions are readily defensible in a potential court case. This consideration often results in very conservative medical decisions and, as a result, persistence in delivering interventions (Fielding 1995).

Social elements may also come into play when the attending family members may need to provide an accounting of treatment decisions to other family members. How might relatives, unaware of the entire context, respond to decisions to withhold treatment? These elements become further complicated when the people making the decisions are perceived, correctly or incorrectly, as having some financial interest in the patient. Often this will be the case since the closest relative and care provider will also be the recipient of the estate. One of the people we interviewed illustrated the strained dynamics of such situations. The interviewee stated, "I was stunned when I arrived in Los Angeles at my mother's death bed to have my youngest sister greet me with, 'When are we going to divide up the money and sell the house? I just want my money in cash now.'" Not surprisingly, the sister's stewardship of the patient was called into question.

ORGANIZATIONAL AND CONTEXTUAL FACTORS INFLUENCING PERSISTENCE

Much of contemporary health care delivery takes place in organizational contexts. These contexts are not merely the passive locations of treatment decisions, but instead actively influence decisions in a variety of ways. As in research on other kinds of investment decisions, three strong organization level factors may commonly influence persistence in treatment decisions. Perhaps most major among these is administrative inertia. Inertia may contribute to continuation of medical interventions since, at least initially, most interventions are not palliative in nature. Even assuming the physician and patient have agreed to end aggressive interventions, it may be difficult to get the other parties involved in health care delivery to be informed of, understand, and accept the decision. Changing procedures that are in place or ensuring that deviations from typical routines are followed is not easily accomplished in organizations.

A major contributor to inertia in medical care delivery is the current extreme segmentation of health care professionals. Problems of segmentation and compartmentalization are not restricted to delivering health care services to the terminally ill. These are typical problems in modern organizations, which tend to be organized along specialization and division-of-labor lines instead of along service process lines (Hammer and Champy 1993). In many organizations, this results in long delays and inefficiencies before the organization can successfully meet the needs of the clients. Specialization and segmentation raise the issue of who is responsible

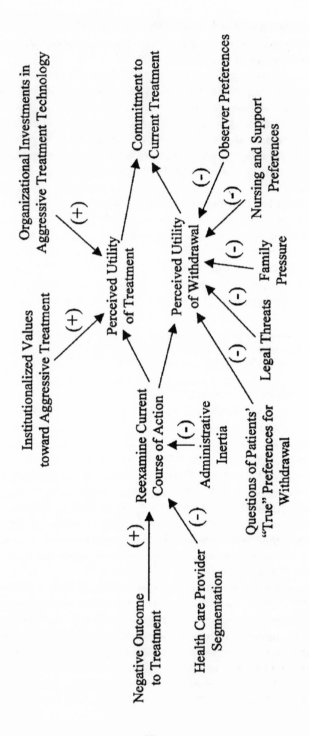

Figure 4. Organizational and Contextual Sources of Continued Intervention

for what. Many doctors or other medical interventionists focus on only one ailment and evaluate their success or failure based on the ailment's response. It is less apparent who is taking the patient's entire well being into account. Thus, in one instance we observed, a patient was diagnosed by his general practitioner as having terminal liver cancer. He was sent to a urologist for treatment of incontinence, who discovered and began to treat the patient's prostate cancer—a treatment that was completely useless since the advanced liver cancer was certain to kill the patient long before the prostate cancer could be of import.

In another of our cases, the hospice that had effectively delivered home care to a patient could no longer visit him when he needed to move into a nursing home facility even though some of the services were nonduplicable and the patient had both the ability and desire to pay for the services. The groups simply did not coordinate with one another. Even in instances in which patients are located in facilities that have excellent palliative case programs, no one is specifically trained in managing the transition from one style of care to the other (Brody, Campbell, Faber-Longerdoen, and Ogle 1997).

Segmentation also appeared to be an issue with one patient's general practitioner. Once he had diagnosed the patient as terminal, the practitioner simply did not know what to do with the patient. His job was treating people and there were not any more treatments that were effective. Yet numerous other health care interactions the patient would be having required his approval and explanation of context. The physician was trained to aggressively intervene to prolong life and improve the quality of life, but he was unprepared and had little capacity to deal with the dying process and the inevitable loss of his patient.

At times decisions or courses of action become institutionalized. By that we mean the projects or programs become tied integrally to the values and purposes of the enterprise. Institutionalized behavior occurs when actions are taken for granted because they are so deeply embedded in the subculture or norm of the group (Scott 1995; Zucker 1977, 1981). In the practice of medicine, fighting death is a core institutionalized value. Thus, for many doctors, the instantaneous reaction to any patient ailment is to treat it as aggressively as possible, given current knowledge, resources, and technology. As Rublee (1994), Bodenheimer (1999), and Bloche (1999) point out, the tendency to overtreatment is probably partially explained by a general tendency to aggressive treatment that is deeply rooted in the culture of American medicine.

Additionally, often contesting death is linked to the basic spiritual values about the sanctity of life. Physicians and nurses may regard this value as taking precedence over the well being of a particular patient at a particular time. Thus, in one instance we observed, a nurse was directed to administer a painkilling injection, which could potentially prove lethal to the suffering terminally ill patient. The nurse evaded these instructions since administering a possibly fatal injection violated her religious beliefs. She felt in a very tangible way that it was a sin, and this belief was far more important than the wishes of patient, family, doctor, or anyone else.

In many major organizational decisions, there also may be a variety of technical side bets that accompany a chosen course of action. These may involve the opening of new plants, the acquisition of new equipment, and the hiring of employees who have specialization in the new arena. It has long been recognized however that the tools used to implement projects often become the driving force for the selection of problems to pursue (March 1978; Weick 1979). Thus, hospitals that have undergone substantial expansions in bed capacity, invested enormous sums in high-tech equipment, and employed dedicated, specialized personnel must find some way to recover these expenditures. Physicians may consciously or subconsciously feel pressure to bring these idle resources to bear on the cases in front of them, particularly in ambiguous circumstances.

In many investment decision contexts, the views of different parties about the desirability of persistence varies. Usually projects are terminated when the correlation of forces favoring withdrawal comes to outweigh the set of forces urging persistence. One of the unique aspects of decisions about end-state medical care is the virtual veto power granted to almost anyone over the decision to discontinue treatment. Obviously the patient's wishes should usually be determinate. Yet even when terminally ill patients repeatedly express their desire to die, psychiatrists point out that this desire is often related to clinical depression, which can potentially be treated. Thus some observers claim that patients' expression of a desire for death are often inherently transient and should be given very limited weight (Chochinov et al. 1995). Discontinuance of interventions may be effectively blocked by doctors, the legal establishment, family, friends, nurses, or at times even complete strangers.

One of the major constituencies influencing health care delivery to terminally ill patients is the legal establishment. The issue of whether and under what circumstances patients have a "right-to-die" is continuing to be debated in the court system. At this point in time, euthanasia is illegal in all states, and assisting suicide is illegal in most. While private physicians will acknowledge hastening patients' demise, it is very uncommon to publicly admit doing so (for an exception, see Quill and Bennett 1992; Quill 1993). Yet questionnaire investigations show that many physicians have received requests for euthanasia and assisted suicide and that, in about a quarter of such cases, assistance has been provided (Angell 1996). How the threat of legal action influences physician behavior is difficult to predict. Qualitative studies find that physicians do not respond in a uniform manner to encroachments into their decision-making domain. Some resist, some adapt to, and some accommodate the various attempts to infringe on their autonomy (Hoff and McCaffrey 1996).

Families may also have a major influence on end-state health care delivery. The conventional wisdom is the families and friends provide valuable social support that augments the individual's well being. However, as Ell (1996) points out, "families are not merely static resource banks from which a seriously ill member withdraws social supports. Families are also potential sources of stress, and negative social exchanges occur in all families." So, for example, the death of a parent may be seen in some instances as the opportunity to rewrite the script of one's childhood. There may be competition among the children to demonstrate commitment to the parent's life—even if such a prolongation meant more financial expenditure and physical suffering for the patient in many cases. In one instance we observed, relatives appeared to view the continued provision of treatments to a patient who was unable to order them withheld as a signal of the fact that they themselves would always want to receive a treatment.

In another instance we encountered, a sister-in-law of a terminally ill patient was insistent that everything possible be done to prolong the clearly terminal patient's life—otherwise who would take care of her car? Such a level of selfishness may be hard to imagine, but these are the realities one encounters when you leave behind abstract discussions and examine decision contexts that exist in the real world.

Other health care providers may also be influential constituencies. One means of insuring a focus on palliative care is for the dying patient to seek treatment in a hospice (Gabel, Hurst, and Hunt 1998). Currently, however, only 17% of dying patients are served by hospices. For many people, home care is virtually impossible, either because they live alone or because their family members work and cannot be home 24 hours a day to provide support. Instead, most Americans die either in hospitals (61%) or nursing homes (17%)(Cassel and Vladeck 1996). In these instances, the nursing staff may come to be a central constituency involved in justifying the selected course of action. In one nursing home where we acted as observers, a nurse clearly derived much of her meaning in life from providing care to terminally ill patients. They grew increasingly dependent on her—a dependence that she found quite psychologically rewarding. She, as well as some of her fellow nurses, was very reluctant to withdraw life-prolonging treatment despite clear instructions from the patient, the family, and the (occasionally) attending physician.

Furthermore, in those contexts in which the normal focus of activities is on prolonging life, decisions to withhold care typically are not limited to a single occasion or discussion. Thus, in one instance we encountered, long after the basic decision had been made among the doctor, patient, and family to do nothing further to prolong the patient's life, numerous additional issue-by-issue discussions arose with the nursing staff. These included a dispute over whether the patient should receive blood transfusions as his red blood cell count declined as well as discussions as to whether to give the patient vitamin supplements such as iron. The emotional strain of these interactions on the patient's family and on the patient was considerable. Often they took place over the terminally ill patient's bed with the nurse or nutritionist stating in an accusatory tone, "You know he will die if he doesn't get these."

In still other cases, the external constituent may have no logical link to the patient. At one nursing home there was a frequent visitor who had a penchant for "adopting" terminal patients. The visitor was highly critical in public of any actions of the staff that involved doing less than everything possible to prolong the patient's life.

FACTORS INFLUENCING PERSISTENCE OVER TIME

As we have seen, the initial decision to begin a medical intervention may reflect a rational calculus focusing on the patient, the ailment, and the physician. As the patient fails to respond positively to treatment, the response is not invariably to halt even painful, costly, and intrusive interventions. A wide variety of psychological and social factors may influence some patients and physicians to continue with treatments (Hibbard, Jewett, Englemann, and Turner 1998). As the patient's health and competence fail, inevitably the decision making power about interventions shifts to those around him or her. Often at this point, the primary physician also withdraws psychologically and at times physically as well. Then organizational and constituency influences may dominate the decision as to whether to continue intervening. One can think of the ghoulish demise of some famous political leaders such as Spain's Franco, or China's Deng, who were kept "alive" for years to serve the purposes of their followers. Yet these examples may only be extreme instances of a relatively common occurrence among less famous individuals.

The above analysis makes no claims of being exhaustive in terms of identifying the forces contributing to continued medical interventions. Rather it reflects the real variables that we encountered operating during our data gathering with terminally ill patients, their families, and health care providers as they interacted in a variety of organizational settings. While the exact representativeness of these experiences is open to question, the data clearly underline the complexity and dynamic nature of treatment interventions. We firmly believe that only models that capture both the range of the stakeholders and the longitudinal unfolding of the process can accurately provide policy guidance.

TOWARD MANAGING EXCESSIVE
MEDICAL INTERVENTIONS

The last decade has seen a slow but steadily increasing public demand for change in the American way of death. Part of the reason for the slow development of outcry against the current American way of dying is post-hoc (pragmatic) rationalization (Gilbert et al. 1998). Just as there is a tendency not to speak ill of the dead, there is a tendency not to speak ill of how the dying person died. Most people intimately involved in the process want to forget about these days, particularly any unpleasant aspects, as quickly as possible. As a consequence, family myths and stories surround the dying or dead family

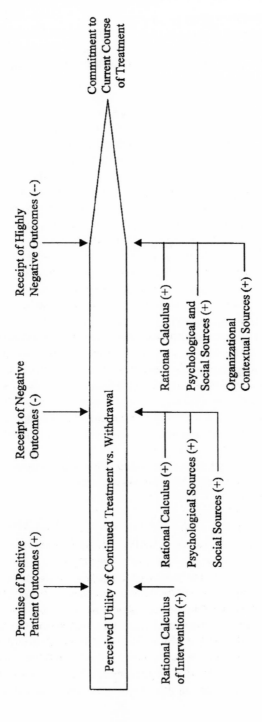

Figure 5. Sources of Intervention Persistence Over Time

member emphasize the pleasant and deny much of the harsh reality and disturbing memories. The survivors often attempt to bury unpleasant memories and painful events with the deceased. The stores have been "laundered." Thus we tend to rationalize and say, "It wasn't so bad after all" or "What's a few weeks of suffering in a life of over 80 years." Yet, for many of us, the memories of these unhappy experiences are difficult to rationalize or suppress. Surely there must be better solutions and more satisfactory resolutions for ourselves and our loved ones.

A variety of individual suggestions has recently been offered for curbing the current tendency to excessive interventions. Enthoven (1993) and Welch and Miller (1994) suggest that a simple alteration in economic incentives will shift the tendency toward less interventions. While economic incentives undoubtedly play a role in health care delivery decisions, we feel that they are far from the only, or even most important, factor. As our analysis has shown, there is a complex array of rational calculus, psychological, social, organizational, and contextual factors, all of which influence treatment decisions.

Observers such as Nuland (1994) believe the solution is closer doctor–patient consultation. So, with clearly terminal patients, for instance, Nuland suggests working with the patient to redefine hope in his or her particular context. This might involve the patient living until a particular important event or having the patient expire only after certain critical matters have been attended to or accomplished. This may well lessen both the patient's and physician's inclination to seek excessive treatment, but it assumes time and effort on the part of the physician and ignores many other factors we have found to influence interventions.

Others unhappy with the current American way of death are pushing for legislation legalizing assisted suicide. It should be recognized that actually hastening patient death pertains to relatively few individuals compared to those where the issue is simply curtailing excessive interventions to prolong life. There is also a wide range of practical problems with such laws, including surveys showing that over half the doctor respondents stated they would not even know what to prescribe in order to assist a suicide.

Over the last twenty years, a variety of mechanisms has been formulated for countering excessive persistence in investment decisions (for a review, see Ross and Staw 1991). Techniques have been developed to assist decision makers to consider a broader range of factors as they engage in the rational calculus of intervention. A variety of procedures such as limit-setting and self-assessment devices have been found to counter psychological forces leading to excessive persistence. Threat-reduction procedures and independent review groups had been proposed to curb some of the social pressures favoring continued commitment. Progress has also been made in developing ways to exit from long developing or late-stage escalation episodes. While a detailed consideration of how these mechanisms might be utilized or adapted to serve in medical decision making is outside the scope of this article, a clear implication of this prior work is that, to effectively curb the tendency for continued interventions, the broad range of factors contributing to excessive persistence must be both understood and addressed.

We believe that no single solution or intervention is likely to be able to remedy problems that are this deep or complex. In the business arena, deciding when to withdraw from failing courses of action is one of the most difficult decisions managers and organizations face. Often these decisions are handled badly resulting in major financial losses. This is true despite decades of training managers to engage in structured cost/benefit analysis and having an entire subdiscipline of management (the area of accounting) geared toward providing data to aid in such decisions. Top managers are selected by organizations, socialized to make "good" decisions about continuing versus withdrawing, and are rewarded based on their ability to implement these decisions. Still the behavioral forces arrayed often lead to excessive persistence. During our analysis of medical decision making for the terminally ill, we have found that most of the behavioral determinants found in management decision contexts also manifest themselves in health care delivery. However, the medical contexts appear to us to be even more complex—and the culture and background of physicians have prepared them far less well to make ideal continue/withdraw decisions.

The medical profession's legal and economic privileges are granted by the public in the expectation that physicians have technical knowledge about medicine and will use that knowledge in the best interests of patients. If physicians and the medical establishment cannot better address the issues surrounding health care delivery to the terminally ill, they will be vulnerable to challenges from contending political and economic groups (Blumenthal 1996).

The medical profession in the United States has responded to social challenges before. The Flexner report led to a major overhaul of medical education in the United States, one that would greatly benefit both the general public and the medical profession. We believe it is time for a major review and revision of how the medical profession provides medical interventions for the terminally ill. The Institute of Medicine has made a major step in this direction through its Committee on Care at the End of Life. In providing a well needed assessment of the western way of dying, the report suggests that "we can do much more to relieve suffering, respect personal dignity, and provide opportunities for people to find meaning in life's conclusion" (Field and Cassel 1997:V). While the committee makes a number of useful suggestions to these ends and calls for more medical education, it does not provide an analytical framework for understanding the forces propelling medicine to increase intervention during this vulnerable period. A logical outgrowth of such a review and development of such an analytical framework would, we believe, be the establishment of a new medical specialty and new forms of analysis.

The specialty will need to have a high level of technical expertise in medicine but with a strong focus on palliative care and ethics. Training must include an understanding of the psychological, social, organizational, and contextual forces surrounding treatment delivery. Organization structures need to be created that support those specialists as they carry out their roles; the medical establishment must both encourage a debate and structure a new majority consensus on how end state care will be administered. Ironically, as the level of technical expertise in medicine continues to rise rapidly, responsive treatment decisions may need to increasingly be made on a nontechnical calculus. Rather, physicians and the larger community of health care providers need a better understanding of the totality of forces and constituents influencing treatment delivery. As this broader perspective is adopted, newly developed medical technologies are more likely to be the servants than the masters of our society's attempts to insure the health of its members.

REFERENCES

Aaron, H.J. 1994. "Thinking Straight About Medical Costs." *Health Affairs* 13:8-13.

Angell, M. 1996. "Euthanasia in the Netherlands—Good News or Bad?" *The New England Journal of Medicine* 335:1676-1678.

Arkes, H.F. and C. Blumer. 1985. "The Psychology of Sunk Costs." *Organizational Behavior and Human Decision Processes* 35:124-140.

Bandura, A. 1977. *Social Learning Theory.* Englewood Cliffs, NJ.: Prentice Hall.

Barbour, A. 1995. *Caring for Patients: A Critique of the Medical Model.* Palo Alto, CA: Stanford University Press.

Bateman, T. 1983. "Resource Allocation After Success and Failure: The Roles of Attributions of Powerful Others and Probabilities of Future Success." Working paper. Department of Management, Texas A&M University, College Station, TX.

Bloche, M.G. 1999. "Clinical Loyalties and the Social Purposes of Medicine." *Journal of the American Medical Association* 281: 268-274.

Bobocel, R.D. and J.P. Meyer. 1994. "Escalating Commitment to a Failing Course of Action: Separating the Roles of Choice and Justification." *Journal of Applied Psychology* 79:360-363.

Blumenthal, D. 1996. "Part 1: Quality of Care-What Is It?" *The New England Journal of Medicine* 335:891-893.

Bodenheimer, T. 1999. "Disease Management- Promises and Pitfalls." *The New England Journal of Medicine* 340: 1202-1205.

Brockner, J. and J.Z. Rubin. 1985. *Entrapment in Escalating Conflicts.* New York: Springer-Verlag.

Brody, H., M.L. Campbell, K. Faber-Longerdoen, and K.S. Ogle. 1997. "Withdrawing Intensive Life-Sustaining Treatment-Recommendations for Compassionate Clinical Management." *The New England Journal of Medicine* 336:652-657.

Cassel, C.K. and B.C. Vladeck. 1996. "CD-9 Code for Palliative or Terminal Care". *The New England Journal of Medicine* 335:1232-1233.

Charmaz, K. 1991. *Good Days, Bad Days: The Self in Chronic Illness and Time.* New Brunswick, NJ: Rutgers University Press.

Chochinov, H.M., K.G. Wilson, M. Enns, N. Mowchun, S. Lander, M. Levitt, and J.J. Clinch. 1995. "Desire for Death in the Terminally Ill." *American Journal of Psychiatry* 152:1185-1191.

Conlon, E.J. and J.M. Parks. 1987. "Information Requests in the Context of Escalation. *Journal of Applied Psychology* 72:344-350.

Danis, M., J. Garrett, R. Harris, and D.L. Patrick. 1994. "Stability of Choices about Life-Sustaining Treatments." *American College of Physicians* 120:567-573.

Danis, M., E. Mutran, J.M. Garrett, S.C. Stearns, R.T. Slifkin, L. Hanson, J.F. Williams, and L.R. Churchill. 1996. "A Prospective Study of the Impact of Patient Preferences on Life-Sustaining Treatment and Hospital Cost." *Critical Care Medicine* 24:1811-1817.

Eddy, D. 1993. "Three Battles to Watch in the 1990s." *Journal of American Medical Association* 273:520-526.

Ell, K. 1996. "Social Networks, Social Support, and Coping with Serious Illness: The Family Connection." *Social Science and Medicine* 42:173-183.

Emanuel, E.J. and L.L. Emanuel. 1994. "The Economics of Dying: The Illusion of Cost Savings at the End of Life." *The New England Journal of Medicine* 330:540-544.

Enthoven, A.C. 1993. "Why Managed Care has Failed to Contain Health Costs." *Health Affairs* 12:27-43.

Field, M.J. and C.K. Cassel (Eds.) 1997. *Approaching Death: Improving Care at the End of Life.* Washington, D.C.: National Academy Press.

Fielding, S.L. 1995. "Changing Medical Practice and Medical Malpractice Claims." *Social Problems* 42:38-55.

Fox, F.V. and B.M. Staw. 1979. "The Trapped Administrator: Effects of Job Insecurity and Policy Resistance Upon Commitment to a Course of Action." *Administrative Science Quarterly* 24:449-471.

Fukkara, A., H. Tazewa, H. Nakajima, and M. Adochi. 1995. "Do Not Resuscitate Orders at a Teaching Hospital in Japan." *The New England Journal of Medicine* 333:805-808.

Gabel, J. R., K. M. Hurst, and K. Hunt. 1998. "Health Benefits for the Terminally Ill: Reality and Perception." *Health Affairs* 17:120-127.

Gattinoni, L., L. Brazzi, P. Pelosi, R. Latini, G. Tognoni, A. Pesenti, and R. Fumagalli. 1995. "A Trial of Goal-Oriented Hemodynamic Therapy in Critically Ill Patients." *The New England Journal of Medicine* 333:1025-1032.

Gilbert, D.T., E.C. Pinel, T.D. Wilson, S.J. Blumberg, and T.P. Wheatley. 1998. "Immune Neglect: A Source of Durability Bias in Affective Forecasting." *Journal of Personality and Social Psychology* 75:617-638.

Goltz, S.M. 1993. "Examining the Joint Rates of Responsibility and Reinforcement History in Recommitment." *Decision Sciences* 24:977-994.

Goodman, P.S., M. Bazerman, and E. Conlon. 1980. "Institutionalization of Planned Organization Change." Pp. 215-246 in *Research in Organizational Behavior.* vol. 2. Edited by B.M. Staw and L.L. Cummings. Greenwich, CT: JAI Press.

Hammer, M. and J. Champy. 1993. *Reengineering the Corporation.* New York: Harper-Collins.

Haug, M.R. 1990. "The Interplay of Formal and Informal Health Care: Theoretical Issues and Research Needs." *Advances in Medical Sociology* 1:207-235.

Hibbard, J.H., J.J. Jewett, S. Englemann and T. Turner. 1998. "Can Medicare Beneficiaries Make Informed Choices?" *Health Affairs* 17: 181-193.

Hoff, T.J. and D.P. McCaffrey. 1996. "Adopting, Resisting, and Negotiating." *Work and Occupations* 23:165-189.

Kuntz, T. 1998. "Death Be Not Unpublishable: The Literature of Good Grief." *The New York Times* November 29, p. 7.

McCain, B.E. 1986. "Continuing Investments Under Conditions of Failure: A Laboratory Study of the Limits of Escalation." *Journal of Applied Psychology* 71:280-284.

March, J.G. 1978. "Bounded Rationality, Ambiguity, and the Engineering of Choice." *Bell Journal of Economics* 9:587-608.

Miller, F.G., T.E. Quill, H. Brody, J.C. Fletcher, L.O. Gostin, and D.E. Miller. 1994. "Regulating Physician-Assisted Death." *The New England Journal of Medicine* 331:199-123.

Nisbett, R. and L. Ross. 1980. *Human Inference: Strategies and Shortcoming of Social Judgement.* Englewood Cliffs, NJ, Prentice-Hall.

Northcraft, G.B. and G. Wolf. 1984. "Dollars, Sense, and Sunk Costs: A Life Cycle Model of Resource Allocation Decisions." *Academy of Management Review* 9:225-234.

Nuland, S.B. 1994. *How We Die: Reflections on Life's Final Chapter.* New York: Knopf.

Orentlicher, D. 1996. "The Legalization of Physician-Assisted Suicide." *The New England Journal of Medicine* 335:663-667.

Pfeffer, J. 1981. *Power in Organizations.* Marshfield, MA: Pittman.

Poplin, C. 1996. "Mismanaged Care." *The Wilson Quarterly* 20:12-24.

Quill, T.E. and N.M. Bennett. 1992. "The Effects of a Hospital Policy and State Legislation on Resuscitation Orders for Geriatric Patients." *Annals of Internal Medicine* 152:569-572.

Quill, T.E. 1993. "Doctor, I want to die. Will you help me?" *Journal of the American Medical Association* 270: 870-873.

Ross, J. and G.L. Albrecht. 1995. "Escalation in Health Care Delivery: Contextual Effects in Physician Diagnosis and Treatment Decisions." *Research in the Sociology of Health Care* 12:241-260.

Ross, J. and B.M. Staw. 1991. "Managing Escalation Processes in Organizations." *Journal of Managerial Issues* 3:19-34.

Ross, J. and B.M. Staw. 1993. "Organizational Escalation and Exit: Lessons from the Shoreham Nuclear Power Plant." *Academy of Management Journal* 36:701-732.

Rublee, D.A. 1994. "Medical Technology in Canada, Germany and the United States: An Update." *Health Affairs* 13:113-117.

Schneiderman, L.J. and N.S. Jecker. 1995. *Wrong Medicine: Doctors, Patients, and Futile Treatment.* Baltimore: The Johns Hopkins University Press.

Scitovsky, A.A. 1994. "The High Cost of Dying Revisited." *The Milbank Quarterly* 72:561-591.

Scott, W.R. 1995. *Institutions and Organizations.* Thousand Oaks, CA: Sage.

Sharpe, V.A. and A.I. Faden. 1996. "Appropriateness in Patient Care: A New Conceptual Framework." *The Milbank Quarterly* 74:115-138.

Staw, B.M. 1976. "Knee-Deep in the Big Muddy: A Study of Escalating Commitment to a Chosen Course of Action." *Organizational Behavior and Human Performance* 16:27-44.

Staw, B.M. and J. Ross. 1980. "Commitment in an Experimenting Society: An Experiment on the Attribution of Leadership from Administrative Scenarios." *Journal of Applied Psychology* 65:249-260.

_____. 1989. "Understanding Behavior in Escalation Situations." *Science* 246:216-220.

Support Principal Investigators. 1995. "A Controlled Trial to Improve Care for Seriously Ill Hospitalized Patients." *Journal of the American Medical Association* 274:1591-1598.

Swanson, J.W. and S.V. McCrary. 1996. "Medical Futility Decisions and Physicians' Legal Defensiveness: The Impact of Anticipated Conflict in Thresholds for End-of-Life Treatment." *Social Science and Medicine* 42:125-132.

Weeks, J.C., E.F. Cook, S.J. O'Day, L.M. Peterson, N. Wenger, D. Reding, F.E. Harrell, P. Kussin, N.V. Dawson, A.F. Conners, J. Lynn, and R.S. Phillips. 1998. "Relationship between Cancer Patients' Prediction of Prognosis and Their Treatment Preferences." *Journal of the American Medical Association* 279: 1709-1714.

Weick, K.E. 1979. *The Social Psychology of Organizing,* (2nd ed.). Redding, MA: Addison-Wesley.

Weiss, C.H. and M.G. Bucuvalas. 1980. "Truth Tests and Utility Tests: Decision Makers' Frame of Reference." *American Sociological Review* 45:302-313.

Welch, W.P. and M.E. Miller. 1994. "Proposals to Control High-Cost Hospital Medical Staffs." Health Affairs 13:42-57.

Wolinsky, F.E. 1988. *The Sociology of Health,* (2nd ed.). Belmont, CA: Wadsworth.

Zucker, L.G. 1977. "The Role of Institutionalization in Cultural Persistence." *American Sociological Review* 42:726-743.

_____. 1981. "Organizations as Institutions." Pp. 1-47 in S.L. Bacharach (ed.) *Perspectives in Organizational Sociology: Theory and Research.* Greenwich, CT: JAI Press.

OLDER ADULTS USE OF PRIMARY CARE PHYSICIANS FOR MEMORY-RELATED PROBLEMS

Neale R. Chumbler, Marisue Cody, Cornelia K. Beck, and Brenda M. Booth

ABSTRACT

Older adults with memory problems (especially those diagnosed with dementia) consume a great number of health services. However, little is known about the special needs of memory-impaired, noninstitutionalized older adults who seek treatment from primary care physicians (PCPs) for memory problems. The study outlined in this chapter investigated the interrelationships among level of education, residence, physical and mental status, and the use of PCPs for memory problems. A probability-based (random-digit dialing) survey of households with individuals more than 60 years of age occurred in six Southern states: Alabama, Arkansas, Georgia, Louisiana, Mississippi, and Tennessee (N = 1,368). We developed two multivariate logistic regression models. The first estimated the additive effects of sociodemographic and health status characteristics on the likelihood of service utilization. The second model inserted two multiplicative terms: the "instrumental

Research in the Sociology of Health Care, Volume 17, pages 31-44.
Copyright © 2000 by JAI Press Inc.
ISBN: 0-7623-0644-0

activities of daily living (IADL)" by "memory impairment" and the "low education" by "memory impairment" interaction terms. Three main findings emerged from the analyses: (1) rural/urban differences in use of PCPs approached statistical significance, (2) for lower educated older adults, the odds of using a PCP declined steadily as their number of memory impairments increased, and (3) as older adults experienced increases in IADL limitations and memory impairments, their likelihood of service use declined. The chapter ends with a discussion on program planning, policy, and practice initiatives in light of the findings.

INTRODUCTION

Substantial changes in cognitive function are among the major features of aging (Barberger-Gateau, Commenges, Gagnon, Letenneur, Sauvel, and Dartigues 1992a). One important change, the loss of memory, interferes with social functioning and psychological well being. Older adults with memory impairment have an increased rate of morbidity and mortality compared to memory-intact older adults (Coward, Netzer, and Peek 1996; Ganguli, Seaberg, Belle, Fischer, and Kuller 1993; Keefover et al. 1996; Liu, LaCroix, and White 1990). When memory-impaired older adults use services, they usually go to primary care physicians (PCPs), who may not recognize cognitive impairment or consider its effects while planning treatment. Thus, memory impairment often goes undetected until the person becomes severely impaired (Buckwalter, Russell, and Hall 1994). The PCPs have low referral rates to specialists (e.g., geriatricians, neurologists, and psychologists). Thus, when PCPs finally do make referrals, these patients may have deteriorated too much to respond to management and treatment (Barberger-Gateau et al. 1992a; Cohen and Cairl 1996; Colenda and van Dooren 1993; Goldstein 1994; Larson 1998).

A dearth of information exists in the literature to guide our understanding of the sociodemographic and health status factors influencing older adults' use of PCPs for memory problems. The present study examined the interrelationships among level of education, residence, physical and memory status, and PCP use for memory problems. A better understanding of these relationships could assist policy makers and planners when targeting populations for more complex cognitive screening and detection of dementia in the future.

At least two reasons justify the need for this type of research: (1) we lack knowledge about whether community-dwelling memory-impaired older adults use PCPs in the same manner and to the same extent as their memory-intact peers use them; (2) with the advent of managed care, PCPs will continue to play a prominent role in treating older adults with memory impairment. Recent guidelines have targeted PCPs to improve detection and diagnosis of dementia (Costa, Williams, and Somerfield 1996; Cummings et al. 1997). Early recognition of cognitive impairment now offers substantial benefits to patients and families

(Whitehouse and Reidenbach 1997). When the cause is reversible, early treatment becomes permanent. When the cause is irreversible, an accurate early diagnosis gives patients and family members time to prepare for future medical, financial, and legal challenges. Unrecognized dementia puts patients at risk for numerous complications, including medication errors, fragmented health care, and delirium. Unrecognized dementia can lead to financial disaster; accidents on the road, in the home, and at work; and unnecessary interpersonal conflicts and alienation from family and friends (McCarten 1997). While no current therapy can reverse progressive cognitive decline, several pharmacologic agents and psychosocial techniques provide relief for the depression, psychosis, and agitation often associated with dementia, and pharmacotherapy may produce shortterm cognitive improvement in many patients. Family-focused support interventions are moderately effective and may delay nursing home placement (Rabins 1996). Thus, early detection, treatment, and appropriate referrals could reduce suffering and costs.

The following discussion has three main sections. The first section consists of a literature review to derive study hypotheses. The second section features a presentation of the sample and measures and a brief introduction of the analytic procedures. The third section reviews and discusses survey results from older adults in six Southern states.

LITERATURE REVIEW

In general, rural residence can stand for a proxy indicator of many sociodemographic and cultural factors (Booth, Ross, and Rost 1999). Older adults in rural areas are generally economically disadvantaged, have more serious and severe health problems, and usually are medically underserved (Gesler and Ricketts 1992). A paucity of information exists on the likelihood of rural older adults using PCPs for treatment of memory problems. Previous studies have found that, compared to urban older adults, rural older adults have greater physical disability, more chronic health conditions, and increased vulnerability. Furthermore, specific risk factors such as hypertension, stroke, and lower levels of education are more pervasive in rural populations (Keefover et al. 1996; Mainous and Kohrs 1995). From studies of general aging populations, older adults residing in rural areas compared to those in urban areas are less likely to use health services because of their limited number, variety, and accessibility (Ganguli et al. 1993). Furthermore, compared to urban older adults, rural older adults are less likely to be aware of health services and to have the financial resources to pay for them. Such difficulties accessing services could have adverse consequences for the complex and sometimes fragile health care needs of rural persons with dementia and their caregivers (Magilvy 1996).

Level of education attained by older adults consistently correlated with both health service use and the likelihood of dementia. Among samples of community-dwelling older adults, greater educational attainment is strongly correlated with increased physician visits (Glasgow 1995; Himes and Rutrough 1994; Wolinsky and Johnson 1991). Multiple previous community-based studies in diverse populations in various countries have found a consistent relationship between limited educational attainment and the increased likelihood of dementia (Callahan, Hall, Hui, Musick, Unverzagt, and Hendrie 1996; Inouye, Albert, Mohs, Sun, and Berkman 1993; Katzman 1993). Katzman (1993) suggests that greater educational attainment provides for a functional reserve of brain capacity that delays the onset of symptoms of cognitive impairment. Moreover, educational attainment is related to numerous chronic health conditions from its relationship to nutritional habits and health behaviors (e.g., smoking, drinking alcohol, etc). Greater educational attainment, therefore, may serve as a marker for better health, resulting in better cognitive performance on a biological basis (Inouye et al. 1993). Thus, older adults with lower educational attainment and greater likelihood of memory impairment should have a lower likelihood of using a PCP for memory problems.

The levels of physical impairment and memory impairment each display strong and consistent associations with the demand of health service. Research has consistently found that elders with greater physical disability were more likely to use health services (e.g., see Chumbler, Beverly, and Beck 1997; 1998; Mitchell and Krout 1998). Adequate cognitive functioning is required to perform IADLs such as handling finances and following complex medical regimens (Fillenbaum et al. 1988; Herzog and Wallace 1997). Thus, in community-based samples of older adults, IADL dependencies were strongly correlated with diagnoses of cognitive impairment (Barberger-Gateau et al. 1992a; Barberger-Gateau, Chaslerie, Dartigues, Commenges, Gagnon, and Salamon 1992b; Fillenbaum et al. 1988). Even though memory-impaired older adults are more ill and frail than the general population (Ganguli et al. 1993; McCormick et al. 1994), studies have mixed results about whether increased memory impairments lead to increased rates of health service use (Ganguli et al. 1993; Herzog and Wallace 1997; Weiler, Lubben, and Chi 1991). From a large sample of noninstitutionalized persons older than 65 years in a rural community in southwestern Pennsylvania, Ganguli and colleagues (1993) found no significant differences among those with cognitive impairment and those who were cognitively intact on visits to physicians. However, this same study did reveal that one-third of the subjects visited their personal physicians less often than once a year. One could argue that the memory impaired would require more frequent PCP contact than the unimpaired both because of potential medical conditions and disability and poor judgment stemming from cognitive impairment in appropriately using PCPs for memory problems (Ganguli et al. 1993). The literature suggests that the combination of increased physical and cognitive impairment are a major challenge for older adults. It further suggests that the impact of IADL limitations on the likelihood of PCP use for memory problems may change at each level of memory impairment.

Four main study hypotheses have emerged from the discussion: (1) older adults residing in rural areas will have greater physical and memory impairment than their urban counterparts will have, (2) older adults in rural areas will be less likely to use PCPs for memory problems, (3) older adults with less education and more memory impairment will be less likely to use PCPs, and (4) older adults with increased levels of both memory and IADL impairment will be more likely to use PCPs. To develop these hypotheses fully requires a large representative sample of older adults. A discussion of the sample and measures for this study follows.

METHODS

Overview

The larger project from which we derived this study sought to examine whether rural at-risk drinkers were less likely to use services for their alcohol problems. The larger project involved brief screening telephone interviews with a stratified random sample of rural and urban adults to identify a cohort of at-risk drinkers (Booth et al. 1999). The sampling frame for conducting the interviews oversampled rural individuals to ensure that a sufficient number of rural households completed interviews. For the current study, rural residence meant living outside a metropolitan statistical area (MSA). By Census Bureau definitions, MSAs need to contain a city with 50,000 or more inhabitants or contain an urbanized area with at least 50,000 people in a county (or counties) with at least 100,000 people (Booth et al. 1999; McLaughlin and Jensen 1998).

Study Procedures

Details on procedures and methods for the larger project appear elsewhere (Booth et al. 1999). It constituted a probability-based random sample of phone numbers in Alabama, Arkansas, Georgia, Louisiana, Mississippi, and Tennessee using a list-assisted random-digit dialing design. The specified respondent for the screening interview was the household member with the most recent birthday. That respondent participated in a brief survey on health care access (Booth et al. 1999).

For the current study, we selected dyads older than 60 years. That is, we selected only those individuals more than 60 years of age with a spouse living in their household. Respondents answered questions for themselves or for the other person older than 60 years in their household. Thus, in some situations, the person using the service may have been the other older person in the household.

Measures

The independent variables considered included a continuous measure for age. African American ethnicity was coded as a dummy variable, making White the reference category. Educational attainment was dichotomized as lower than a ninth-grade education (subsequently referred to as *low education*) and a ninth-grade education and higher (the reference category). Rural residence meant living outside an MSA (urban residence was the reference category). We assessed physical and cognitive function. To measure physical function, the respondent reported the extent of difficulty performing IADLs in the past month (e.g., telephone use, use of transportation, responsibility for medication intake, and handling finances; higher value=more limitations). To evaluate cognitive function, we employed the four items that exclusively focused on memory, from the seven-item *Memory-Related subscale* of the 24-item *Revised Memory and Behavior Problems Checklist* (Teri et al. 1992). The four items follow: In the past month, did you or someone in your household more than 60 years of age have (1) a problem with asking the same question over and over, (2) trouble remembering recent events (e.g., items in the newspaper or on TV), (3) a problem with losing or misplacing things, and (4) a problem with forgetting what day it is? To determine utilization of PCPs for memory problems, the dependent variable, a single item asked respondents whether they or the other person older than 60 years in their household had visited a PCP for memory problems in the past 12 months.

ANALYTIC PROCEDURES

Weights were calculated to reflect a combination of sampling weights and nonresponse weights as well as to counterbalance the oversampling of phone numbers in non-MSA counties. In addition, the weights modified the geographic distribution of the sample to the population distributions in the six states separately by age, gender, race, and rurality (Booth et al. 1999).

Our analyses consisted of two main steps. First, we employed chi-square analyses (for dichotomous variables) and *t*-tests (for continuous variables) to compare weighted frequencies and weighted means among both rural and urban respondents. Next, we developed two multivariate logistic regression models. In the first model, we estimated the additive effects of the sociodemographic and health status characteristics on the likelihood of service utilization. In the second model, we assessed the multiplicative terms of the IADL impairment–memory impairment and low education–memory impairment interaction terms. The interaction terms were centered on their means to avert multicollinearity problems frequently associated with interaction effects (Aiken and West 1991; McClendon 1994). Additional analyses derived two estimates: The first assessed the influence of low education on service use for those with no memory impairments, one memory impairment, or more than two memory impairments, while the second captured the influence of IADL impairments on service use for those with no memory impairments, one memory impairment, or more than two memory impairments.

In both models, we reported the adjusted odds ratios to indicate the net change in the likelihood of service utilization given a one-unit increase in the independent variable. The Hosmer-Lemeshow goodness-of-fit statistic (Hosmer and Lemeshow 1989) estimated model fit.

RESULTS

Table 1 depicts the weighted descriptive statistics for all study variables. It also compares the basic demographic and health characteristics of rural and urban respondents and indicates several significant differences. There were 736 (53.8%) urban respondents and 632 (46.2%) rural respondents. No significant rural/urban difference emerged in PCP use rates (10.6% vs.11.5%, respectively). However, compared to urban respondents, rural respondents tended to be older, White, and less educated and had a greater number of memory and IADL impairments.

Table 2 contains the results of the logistic regression models designed to evaluate the effects of the sociodemographic and health status characteristic variables on PCP use. Model 1 in Table 2 shows the adjusted odds ratios obtained when only the additive terms were included in the equation. The Model 2 coefficients in Table 2 resulted from the entry of the multiplicative terms.

Table 1. Characteristics of the Sample by Area of Residence

Variables: Sociodemographic Characteristics	Percentages/Means			Significance x^2 or t-test
	Urban (736)	Rural (632)	Total (1368)	
Age:	69.23	69.44	69.33	< .001
(Range: 60–91)	6.43	7.12	6.75	
Ethnicity				
White	78.8%	83.2%	80.8%	.041
Non-White	21.2%	16.8%	19.2%	
Education				
No formal education–8th grade	15.8%	22.4%	18.8%	.001
≥ 9th grade	84.2%	77.6%	81.2%	
Memory Impairment	.55	.66	.60	.009
(Range: 0–4)	(.96)	(1.02)	(.99)	
IADL Limitations	.14	.17	.15	.058
(Range: 0–4)	(.48)	(.51)	(.49)	
Primary Care Physician Use	11.5%	10.6%	11.1%	.578

Note: Standard deviations shown in parentheses.

Table 2. Logistic Regression Estimates for Likelihood of
Use of Primary Care Physicians for Memory Problems ($N = 1,368$)

	Model 1		Model 2	
Independent Variable	β	AOR (95% C.I.)	β	AOR (95% C.I.)
African American (vs. white)	.22	1.25 (.78–1.99)	.06	1.06 (.65–1.71)
No formal education to 8th grade[f]	.03	1.04 (.64–1.67)	.27	1.31 (.79–2.17)
Age	.02	1.03 (1.00–1.05)	.02	1.02 (.99–1.05)
Rural Residence	−.30	.74 (.50–1.10)	−.37	.69[b] (.46–1.03)
IADL limitations	.72	2.05[e] (1.48–2.85)	1.63	5.12[e] (3.20–8.20)
Number of Memory impairments	.83	2.29[e] (1.95–2.70)	.96	2.62[e] (2.20-3.11)
(No formal education–8th grade x number of memory impairments)	—	— —	−.72	.49[c] (.27–.89)
(IADL limitations x number of memory impairments)	—	— —	−.47	.63[e] (.52–.75)
Hosmer-Lemeshow Goodness-of-fit (p value)	78.55	(.00)	66.75	(.00)

Note: [a]β = Unstandardized logistic regression coefficient; AOR = Adjusted Odds Ratio; C.I. = confidence interval.
[b]$p = .07$,
[c]$p \leq .05$;
[d]$p \leq .01$;
[e]$p \leq .0001$
[f]Reference category: $\geq$ 9th grade education

Model 1 in Table 2 indicates that, for each additional memory impairment and IADL limitation, the odds of using a PCP for memory problems were 2.29 and 2.05, respectively, times greater, holding all other variables constant. Model 2 in Table 2 indicates that for each additional memory impairment and IADL limitation, the odds of using a PCP for memory problems were 2.62 and 5.12, respectively, times greater, holding all other variables constant. Living in a rural residence was associated with a 31% lower likelihood ($p = .07$) of using a PCP for memory problems, holding all the other variables constant. Furthermore, the data in this model revealed statistically significant interaction effects between number of memory impairments and lower education and number of IADL limitations on the likelihood of PCP use.

We divided the memory impairment measure into three groups: (1) no impairment ($n = 874$); (2) one impairment ($n = 256$); and (3) more than two impairments ($n = 238$). The data in Table 3 indicated that, for lower educated older adults and those with more IADL impairments, the odds of using a PCP for memory problems declined steadily as the number of memory impairments increased.

Table 3.   Odds Ratios for Interaction Terms From
Logistic Regression Analyses of Primary Care Physician Use

	0 Memory Impairments	1 Memory Impairment	≥ 2 Memory Impairments
Education			
No formal education–8th grade	3.18[a]	1.55[b]	.71[b]
≥ 9th grade education	.32[a]	.64[b]	1.42[b]
IADL Limitations			
+1 SD	28.80[a]	1.87[a]	1.60[b]
–1 SD	.04[a]	.53[a]	.63[b]

Note: [a]$p \leq .05$
        [b]$p \geq .05$

DISCUSSION

This study in six southern states presented some new information on older adults' utilization of PCPs for memory problems. We sought to identify some character-istics of older adults that researchers could target for more complex cognitive screening and detection of dementia in the future. Four major conclusions were drawn from these data:

1.   Consistent with Hypothesis 1 is our finding that rural older adults had more memory and IADL impairments than urban older adults did. These significant bivariate associations along with the findings that rural older adults had attained less education supported past research (Mainous and Kohrs 1995).

2.   Despite this disadvantage and inconsistently with Hypothesis 2, rural elders were almost as likely as urban elders to have used PCPs for memory prob-lems. However, we should point out that, in the multivariate model that included the two multiplicative terms, the inverse association between rural residence and use of PCPs approached statistical significance. The measure we employed to operationalize rural residence, the Census Bureau definition of living outside an MSA, may partially explain the absence of an independent rural effect. Our rural sample certainly included residents of counties with relatively suburban qualities and access to many urban services as well as residents of locations with scant pop-ulations and few health care resources (Booth et al. 1999). There are substantial differences in social milieu in the southern states covered by our study, unques-tionably even within single states. Future research should use expanded defini-tions of residence to distinguish between areas adjacent and not adjacent to MSAs and examine the unique mix of prevention activities, treatment planning, and policy development for defined rural areas.

3.   Consistent with Hypothesis 3 was our finding that support for the expecta-tion that the combination of lower education attainment and higher levels of

memory impairment were associated with a decreased likelihood of PCP use for memory problems.

4. Incompatible with Hypothesis 4 was the increased levels of both memory impairment and IADL impairment being associated with a lower likelihood of PCP use, a finding that went in the *opposite* direction of our hypothesis. These observations make us wonder where memory-impaired older adults seek services as their need increases and also call for a more careful assessment of how medical and cognitive comorbidity, which exacerbates diminished quality of life and perpetuates premature institutionalization, contributes to the need for PCPs (Colenda and van Dooren 1993).

Our findings could have various implications for program planning, practice initiatives, and screening efforts. PCPs must watch for early symptoms of dementia and recognize that many patients seek out medical treatment for a reason other than cognitive difficulty (Small et al. 1997). Patients delay reporting symptoms to medical professionals because they attribute cognitive losses to normal aging. Family members delay reporting symptoms out of respect for their parents or grandparents and because of the negligible effect of the dementing illness on family life and the family economy in the early stages (Iliffe 1997). A study out of the Michigan Alzheimer's Disease Center reported that families found it difficult to locate a physician with expertise in dementing illnesses, the physician often failed to disclose the diagnosis in an informative and compassionate manner, families wanted their family physician involved in the diagnostic and assessment process, and families needed referrals to community-based services (Connell and Gallant 1996).

Our results are important for two reasons: (1) we collected information on non-users of PCPs—information that allowed us to especially identify some access issues—and (2) we collected information from a wide geographic area ranging from urban (i.e., Atlanta and New Orleans) to rural (i.e., Mississippi Delta) regions with a broad variety of residents. The variety may be associated with differential PCP-use patterns as well as higher prevalence of physical and memory impairment. Furthermore, PCPs and mental health providers have traditionally underserved the rural south with its high concentration of poverty, especially among rural older adults living in the Mississippi Delta region of Arkansas and Mississippi. Residents of these two states rank at the bottom of most measures of poverty and health status (Chumbler et al. 1997; Glasgow and Brown 1998).

Some important caveats of the present study need to be mentioned. First of all, we relied on self-reported memory problems. Furthermore, the self-report measure we employed was originally developed in a self-administered, paper-and-pencil format, not over the telephone by an unknown interviewer. Therefore, the measures' reliability and validity remain unknown. We do not intend for others to view these four items solely to diagnose dementia. These four items, plus telephone screenings for dementia, should accompany a complete clinical evaluation. Respondents may have underestimated the information of physical and memory impairment because of fears concerning confidentiality and anonymity of the data. Even with this known shortcoming, our self-report measure did contain a range of limitations. This allowed us to count how many memory impairments were correlated with use of services. Furthermore, cognitively impaired seniors can accurately report use or nonuse of a specific service (Fox 1997).

Second of all, our protocol specifically collected information on those individuals more than 60 years of age with one additional person more than 60 years of age living in their household. Proxy reports, although imperfect, are often the only origin of self-report data accessible and are valuable data sources for measuring health service utilization. Also, proxies' ratings of cognitive status typically correspond well with mental status test results (Magaziner 1997).

In respect to response bias, proxies tend to rate cognitive status better than mental status exams disclose (Magaziner 1997). We only collected minimal information from elders or their proxies. Additional information could have provided valuable insight to salient access issues such as transportation to services and the knowledge and awareness of services. Future research must cautiously document use of proxies and the possible error their use presents. Additional research must contrast proxy reports with the "gold standard," criterion measures, to achieve greater verification from proxy reports (Magaziner 1997).

In summary, public policy has a responsibility to establish an effective service delivery system for memory-impaired rural older adults. Older adults with memory problems are underserved in the community because PCPs often do not recognize the seriousness of memory impairment and specialized services are rarely available. Expansion of service options would provide valuable information for planners and policy makers to accommodate services in ways that match the needs of memory-impaired older adults and their family caregivers.

ACKNOWLEDGMENTS

The work reported here was supported by a grant (R01AA10372) from the National Institute on Alcohol and Alcoholism to Brenda M. Booth (Principal Investigator). The authors gratefully acknowledge the contributions of Valorie Shue, Andrew Weier, Sue Kalepp, Lorraine Porcellini, and Stacy Kimbrel. Address correspondence to Neale R. Chumbler, Ph.D., Center for Health Services Research, Marshfield Medical Research Foundation, 1000 North Oak Avenue, Marshfield, WI 54449-5790. Telephone: (715) 387-9148; FAX: (715) 389-4788; e-mail: chumblen@mfldclin.edu

REFERENCES

Aiken, L.S., and S.G. West. 1991. *Multiple Regression: Testing and Interpreting Interactions.* Newbury Park, CA: Sage.

Barberger-Gateau, P., D. Commenges, M. Gagnon, L. Letenneur, C. Sauvel, and J. Dartigues. 1992a. "Instrumental Activities of Daily Living as a Screening Tool for Cognitive Impairment and

Dementia in Elderly Community Dwellers." *Journal of the American Geriatrics Society* 40:1129-1134.

Barberger-Gateau, P., A. Chaslerie, J.F. Dartigues, D. Commenges, M. Gagnon, and R. Salamon. 1992b. "Health Measures Correlates in a French Elderly Community Population: The PAQUID Study." *The Journals of Gerontology* 47:S88-S95.

Booth, B.M., R.L. Ross, and K. Rost. 1999. "Rural and Urban Problem Drinkers in Six Southern States." *Substance Use and Misuse* 34: 471-493.

Buckwalter, K.C., D. Russell, and G. Hall. 1994. "Needs, Resources, and Responses of Rural Caregivers of Persons with Alzheimer's Disease." Pp. 301-315 in *Stress Effects of Family Caregivers of Alzheimer's Patients,* edited by E. Light, G. Niederehe, and B.D. Lebowitz. New York: Springer.

Callahan, C.M., K.S. Hall, S.L. Hui, B.S. Musick, F.W. Unverzagt, and H.C. Hendrie 1996. "Relationship of Age, Education, and Occupation With Dementia among a Community-Based Sample of African Americans." *Archives of Neurology* 53:134-140.

Chumbler, N.R., C.J. Beverly, and C.K. Beck. 1998. "Determinants of In-Home Health and Support Service Utilization for Rural Older Adults." *Research in the Sociology of Health Care* 15:205-227.

Chumbler, N.R., C.J. Beverly, and C.K. Beck. 1997. "Rural Older Adults' Likelihood of Receiving a Personal Response System: The Arkansas Medicaid Waiver Program." *Evaluation and Program Planning* 20:117-127.

Cohen, D., and R. Cairl, 1996. Mental Health Care Policy in an Aging Society. Pp. 301-319 in *Mental Health Services: A Public Health Perspective.* edited by B.L. Levin and J. Petrila. New York: Oxford.

Colenda, C.C., and H. van Dooren. 1993. "Opportunities for Improving Community Mental Health Services for Elderly Persons." *Hospital and Community Psychiatry* 44:531-533.

Connell, C.M., and M.P. Gallant. 1996. "Spouse Caregivers' Attitudes toward Obtaining a Diagnosis of a Dementing Illness." *Journal of the American Geriatrics Society* 44: 1003-1009.

Costa, P.T., T.F. Williams, and M.R. Somerfield. 1996. Recognition and Initial Assessment of Alzheimer's Disease and Related Dementias. Clinical Practice Guidelines No. 19. Rockville, MD: U.S. Department of Health and Human Services.

Coward, R.T., J.K. Netzer, and C.W. Peek. 1996. "Obstacles to Creating High-Quality Long-Term Care Services for Rural Elders." Pp. 10-34 in *Long-Term Care for the Rural Elderly,* edited by G.D. Rowles, J.E. Beaulieu, and W.M. Myers. New York: Springer.

Cummings, J.L., J. Boos, B.D. Dickinson, M.G. Hazlewood, L.F. Jarvik, K.A. Matuszewski, and R.C. Mohs. 1997. *Dementia Identification and Assessment: Guidelines for Primary Care Practitioners.* Washington, D.C.: Department of Veterans Affairs.

Fillenbaum, G.G., D.C. Hughes, A. Heyman, L.K. George, and D.G. Blazer. 1988. "Relationship of Health and Demographic Characteristics to Mini-Mental State Examination Score among Community Residents." *Psychological Medicine* 18:719-726.

Fox, P.J. 1997. "Service Use and Cost Outcomes for Persons with Alzheimer Disease." *Alzheimer Disease and Associated Disorders* 11 (Suppl. 6):125-134.

Ganguli, M., E. Seaberg, S. Belle, L. Fischer, and L.H. Kuller. 1993. "Cognitive Impairment and the Use of Health Services in an Elderly Rural Population: The MoVIES Project." *Journal of the American Geriatrics Society* 41:1065-1070.

Gesler, W.M., and T.C. Ricketts. 1992. *Health in Rural North America: The Geography of Health Care Services and Delivery.* New Brunswick, NJ: Rutgers University Press.

Goldstein, M.Z. 1994. "Taking Another Look at the Older Patient and the Mental Health System." *Hospital and Community Psychiatry* 45:117-119.

Glasgow, N. 1995. "Retirement Migration and the Use of Services in Nonmetropolitan Counties." *Rural Sociology* 60:224-243.

Glasgow, N., and D.L. Brown. 1998. "Older, Rural and Poor." Pp. 187-207 in *Aging in Rural Settings: Life Circumstances and Distinctive Features,* edited by R.T. Coward and J.A. Krout. New York: Springer.

Herzog, A.R., and R.B. Wallace. 1997. "Measures of Cognitive Functioning in the AHEAD Study." *The Journals of Gerontology* Series B 52B *(Special Issue)*:37-48.

Himes, C.L., and T.S. Rutrough. 1994. "Differences in the Use of Health Services by Metropolitan and Nonmetropolitan Elderly." *The Journal of Rural Health* 10:80-88.

Hosmer, D.W., and S. Lemeshow. 1989. *Applied Logistic Regression.* New York: Wiley-Interscience.

Iliffe, S. 1997. "Can Delays in the Recognition of Dementia in Primary Care be Avoided?" *Aging and Mental Health* 1:7-10.

Inouye, S.K., M.S. Albert, R. Mohs, K. Sun, and L.F. Berkman. 1993. "Cognitive Performance in a High-Functioning Community-Dwelling Elderly Population." *Journal of Gerontology: Medical Sciences* 48: M146-M151.

Katzman, R. 1993. "Education and the Prevalence of Dementia and Alzheimer's Disease." *Neurology* 43:13-20.

Keefover, R.W., E.D. Rankin, P.M. Keyl, J.C. Wells, J. Martin, and J. Shaw. 1996. "Dementing Illnesses in Rural Populations: The Need for Research and Challenges Confronting Investigators." *The Journal of Rural Health* 12:178-187.

Larson, E.B. 1998. "Management of Alzheimer's Disease in a Primary Care Setting." *American Journal of Geriatric Psychiatry* 6 (2)(Suppl.1):S34-S40.

Liu, I.Y., A.Z. Lacroix, and L.R. White. 1990. "Cognitive Impairment and Mortality: A Study of Possible Confounders." *American Journal of Epidemiology* 41:136-143.

Magaziner, J. 1997. "Use of Proxies to Measure Health and Functional Outcomes in Effectiveness Research in Persons with Alzheimer Disease and Related Disorders." *Alzheimer Disease and Associated Disorders* 11 (Suppl.6):168-174.

Magilvy, J. 1996. "The Role of Rural Home- and Community-Based Services." Pp. 64-84 in *Long-Term Care for the Rural Elderly: New Directions in Services, Research, and Policy,* edited by G. Rowles, J.E. Beaulieu, and W. Myers. New York: Springer.

Mainous, A.G.I., and F.P. Kohrs, 1995. "A Comparison of Health Status between Rural and Urban Adults." *Journal of Community Health* 20:423-431.

McCarten, J.R. 1997. "Recognizing Dementia in the Clinic: Whom to Suspect, Whom to Test." *Geriatrics* 52 (Suppl.2):S17-S21.

McClendon, M.J. 1994. *Multiple Regression and Causal Analysis.* Itasca, IL: F.E. Peacock

McCormick, W.C., W.A. Kukull, G. van Belle, J.D. Bowen, L. Teri, and E.B. Larson. 1994. "Symptom Patterns and Comorbidity in the Early Stages of Alzheimer's Disease." *Journal of the American Geriatrics Society* 42:517-521.

McLaughlin, D.K., and L. Jensen. 1998. The Rural Elderly: A Demographic Portrait. Pp. 15-43 in *Aging in Rural Settings: Life Circumstances and Distinctive Features,* edited by R.T. Coward and J.A. Krout. New York: Springer.

Mitchell, J., and J.A. Krout. 1998. "Discretion and Service Use Among Older Adults: The Behavioral Model Revisited." *The Gerontologist* 38:159-168.

Rabins, P.V. 1996. "Behavioral Disturbances of Dementia: Practical and Conceptual Issues." *International Psychogeriatrics* 8 (Suppl.3):281-283.

Small, G.W., P.V. Rabins, P.P. Barry, N.S. Buckholtz, S.T. DeKosky, S.H. Ferris, S.I. Finkel, L.P. Gwyther, Z.S. Khatchaturian, B.D. Lebowitz, T.D. McRae, J.C. Morris, F. Oakley, L.S. Schneider, J.E. Streim, T. Sunderland, L.A. Teri, and L.A. Tune. 1997. "Diagnosis and Treatment of Alzheimer Disease and Related Disorders: Consensus Statement of the American Association for Geriatric Psychiatry, the Alzheimer's Association, and the American Geriatrics Society." *Journal of the American Medical Association* 278:1363-1371.

Teri, L., P. Truax, R. Logsdon, J. Uomoto, S. Zarit, and P.P. Vitaliano, 1992. "Assessment of Behavioral Problems in Dementia: The Revised Memory and Behavior Problems Checklist." *Psychology and Aging* 7:622-631.

Weiler, P.G., J.E. Lubben, and I. Chi. 1991. "Cognitive Impairment and Hospital Use." *American Journal of Public Health* 81:1153-1157.

Whitehouse, P.J., and F. Reidenbach. 1997. Editorial. "Guidelines for Early Identification of Alzheimer Disease." *Alzheimer Disease and Associated Disorders* 11:61-62.

Wolinsky, F.D., and R.J. Johnson. 1991. "The Use of Health Services by Older Adults." *Journal of Gerontology* 46:S345-S357.

PART II

HEALTH PROFESSIONS AND OCCUPATIONS

HEALTH PROFESSIONAL LICENSURE:
INDIVIDUAL VERSUS INSTITUTIONAL POLICY DEBATE

Elizabeth Furlong and Marlene Wilken

ABSTRACT

Changes in the health care system are threatening the independence and profession-alism of health care providers. One such proposed policy change related to profes-sional licensure is creating a specific challenge for the nursing profession. This article examines how the licensure challenge affects aspects of autonomy, indepen-dence, and integrity of the nursing profession with special emphasis on one group of nurses—that of nurse practitioners. The impact of the PEW Commission's study is analyzed as well as the concern for potential loss of individual licensure for nurses. Nursing groups have responded to the PEW Commission study to prevent such deregulation of the nursing profession.

Research in the Sociology of Health Care, Volume 17, pages 47-61.
Copyright © 2000 by JAI Press Inc.
All rights of reproduction in any form reserved.
ISBN: 0-7623-0644-0

This chapter describes how the autonomy of one health profession, nursing, is being threatened by the changes in the health care delivery system and by recommendations of some policy makers. A background context of the sociology of nursing describes how, in an unprecedented way, the independence, autonomy, and integrity of this health profession is being challenged by a proposed policy that would position licensing and credentialing of health professionals with one's employing organization versus the traditional state profession-specific regulatory board. The impact of the policy recommendations emerging from the PEW Health Professions Commission is emphasized. Furthermore, the effects on one subset of nurses, nurse practitioners, will also be included. Finally, action steps for professionals are listed to ensure continued autonomy of the profession.

The sociological account of the professions of medicine and nursing includes economics, culture, and autonomy. Historically, nursing has experienced recurrent cycles of being nonvalued. The lack of value placed on nursing stems from several factors. Nursing is a predominately female profession where women's work has consistently been undervalued. The core of nursing centers around caring with deep roots in religious orders, dedication, charity, and expectations of service. Caring is at the very root of women's history. The ideology of a woman's sphere supported the notion that female subordination was necessary for societal stability. Nursing was established to be a place where the women could legitimately work outside the home, but because of nursing's association with domesticity and caring it was undervalued.

For many years, the culture of nursing was predominately medically based. Nursing used the medical model as a basis for providing patient care. The medical model is centered on human pathology, that is, the focus of care is on the disease process. Physicians are taught to cure the disease and remain detached from the patient. For nursing, patient care is based on the person's response to the illness and their health. The pathology is only one part of a complex matrix. By using the medical model, nursing was relegated to a minor role, while curing, medicine's main focus, assumed a major role.

The role of autonomy in nursing and medicine is changing. Historically, some would contend that the main focus between the professions of nursing and medicine is about subordination of nursing to medicine. Since the mid-1920s, physicians have enjoyed a status of dominance and most admired profession. The relationship between physician and nurse is patriarchal. Physicians rarely treat nurses as equals. Doctors major form of communication is through doctor's orders rather than collegial discussion based on equity (Sweet and Norman 1995).

Sociological theory is valuable in explaining the evolution of the nurse–doctor relationship in terms of sexual division of labor and gender-role stereotyping. Although there is little research-based literature on the subject of doctor–nurse relations, what is available speaks volumes to the relationship between physician and nurse. Power and status differentials between doctors and nurses were studied by Keddy and colleagues (1986). The study identified the "doctor–nurse game," which is the engagement of stereotypical patterns of interaction based on gender differences in society. Stein (1967), who originally described the "game," indicated that the game ensures that open disagreement is avoided at all cost between the nurse and physician. The nurse learns to show initiative and offer significant advice while appearing to defer passively to the doctor's authority. Nurses use subtle nonverbal and cryptic verbal cues to communicate recommendations that, in retrospect, appear to have been initiated by the physician.

Stein describes medical students as being phobic about making mistakes and thus as developing a defensive belief that they are omnipotent and omniscient. This interferes with them taking orders from nonphysicians. Nurses, who are educated in tightly disciplined curriculums and inculcated with subservience and a fear of independent action, are bound to play the game as a way out of the bind.

Stein (1990) revisited the doctor–nurse game to evaluate changes in his original theory. Relationships between the two professions have improved. More recent studies indicated that, although the game may still be observed, the game occurs less often and nurses are more straight forward about giving advice. Perhaps this is due in part to the fact that nursing has become more prominent. Nursing has progressed from an occupation to a profession. The image of nursing is changing, albeit slowly, from that of a dependent handmaiden to an assertive practitioner. The change in image is due in part due to feminist scholarship and the rethinking about gender roles in professions.

Changes are occurring not only in the sociology of the nursing profession but also in the health care environment. The rapidly changing health care environment may portend changes in the roles, functions, and structure of the various health professions. The environmental pressures, which are facilitating these changes, include the rapidly evolving managed care paradigm, an emphasis on outcome measures in the health system, and a renewed emphasis on desiring competencies in health providers versus the traditional concern with the credentials of health professionals. There are several indicators of these changes. By 2005, it is predicted that 85% of Americans will be enrolled in a managed care plan ("Critical Challenges..." 1995). The increase in managed care created internal shifts in health delivery systems from a not-for-profit paradigm to a for-profit paradigm. In 1985, not-for-profit managed care plans outnumbered for-profit plans by two to one. By 1995, however, the ratio was reversed with twice as many for-profit Health Maintenance Organizations (HMOs) as not-for-profit HMOs. This change in underlying values is having an impact in the delivery of health care ("Nursing's Values Challenged..." 1998). Loss of autonomy—the "proletarianization of the doctor" to use McKinlay's evocative phrase—was powerfully augmented by political and economic factors that led to further undermining of the doctor's decisive role in medicine (Silver 1997). Concurrent with these changes, loss of autonomy and proletarianization, are the concerns that health professionals have about the decreasing quality of care for patients. A national survey of health professionals yielded these results: (1) there is a trend toward declining quality, (2) several national trends in health care are negatively affecting patient care, and (3) understaffing is a concern ("Yet Another Survey" 1998).

In this new for-profit paradigm, there will be increased emphasis on preventive health, outcome measures, population-based health status indicators, primary care versus speciality care, and community-based care. These changes will result in surpluses of some health providers, different educational preparation of health providers, and, potentially, the emergence of a multiskilled worker. Predictions of changes for health care personnel and health organizations include the following: (1) closure of more than half of the nation's hospitals with a loss of 60% of hospital beds and a massive expansion of health care to community settings, (2) surpluses of 150,000 physicians, 300,000 nurses, and 40,000 pharmacists, and (3) the consolidation of over 200 allied health professionals into multiskilled professionals ("Critical Challenges..." 1995). Increased corporate ownership of health professional schools is predicted because it is in the interest of corporate health care organizations to educate and train their own workers, which is analogous to the kind of training that business corporations implement with their new employees. The combination of such pressures calls into question the continued autonomy of health professionals.

The PEW Commission has been a major actor in addressing the pressures and changes in the health policy arena. This private foundation is administered by the Center for the Health Professions at the University of California, San Francisco. Among the Commission's goals is the assistance they provide to health professionals, health professional schools, health care organizations, and public policy makers to respond to the challenges of educating and managing a health care workforce that is competent to meet the needs of the nation. In an effort to meet these goals, the PEW Commission has published numerous studies, many of which evolved into national conferences. These studies have been the focus of many articles in a large variety of health professional journals. The Pew Commission has invested more than $70 million dollars in the past decade to study and advance health care professions education (PEW Health Professions Commission Revived 1997).

The Commission's first study, *Healthy America: Practitioners for 2005*, was released in 1991. In this study, seventeen competencies were identified that health professionals needed to have to work as competent professionals in the year 2005. Health science schools were encouraged to prepare their students with these competencies. Conceptually, the PEW Commission was problem-solving some of the concerns of the changing health care system by focusing on the education of health science professionals. By being competent in these areas, certain challenges of the health care system would then be addressed. The competencies include

1. caring for the community's health
2. expanding access to effective care
3. providing clinically competent care
4. emphasizing primary care
5. participating in coordinated care
6. ensuring cost effective and appropriate care
7. practicing prevention
8. involving patients and families in the decision making process
9. promoting healthy lifestyles
10. assessing and use technology appropriately
11. improving the health care system
12. managing information
13. understanding the role of the physical environment
14. providing counseling on ethical issues
15. accommodating expanded accountability
16. participating in a racially and culturally diverse society
17. continuing to learn

In 1993, the Commission released another study called *Health Professions Education for the Future: Schools in Service to the Nation Reform Strategies.* Following the 1991 and 1993 studies, there was a concern that change would not come fast enough using the strategy of going through the health science schools. Therefore, the Commission released another study in 1995—*Reforming Health Care Workforce Regulations.* This study sparked much reaction in the professional literature and was the impetus for many national conferences initiated by health professional groups in response to this 1995 Report. Conceptually, this study calls for legislative regulatory reform to facilitate role changes of health professionals rather than attempting to accomplish change via the educational route of health science schools. One rationale for this change in strategy is that the vested interests in health professional schools and the bureaucracy of educational institutions precluded them from moving forward in rethinking and changing the roles of health professionals. Some of these proposed regulatory changes are threatening to the traditional autonomy and integrity of health professions. This has created much discussion and concern among health professionals. The question for some health providers is "Will regulatory changes cause a demise of their profession?"

The 1995 study by the PEW Commission was predicated on several concerns about the health care system in the United States. The concerns include rising costs, restrictive managerial and professional flexibility in the use of various health personnel, decreased access to health care for some, an equivocal relationship between current regulatory laws of health professionals and quality of care, and nonaccountability to the public consumer.

The regulatory changes recommended by the PEW Commission include the following ten elements:

1. standardize regulatory terms
2. standardize entry-to-practice requirements
3. remove barriers to the full use of competent health professionals
4. redesign health professional regulatory board structure and function
5. inform the public about practitioner practices
6. collect data on the health professions
7. assess practitioner competence and assure continuing competence
8. reform the professional disciplinary process
9. evaluate regulatory effectiveness
10. understand the organizational context of health professions regulations.

Some health care professionals are concerned about one recommendation in particular. Removing barriers that allow for the full use of competent health professionals may result in diminishing the autonomy of nurses. Nurses have identified that the removal of barriers could replace individual licensure with institutional licensure. In addition, a multiskilled worker may emerge in place of a profession-specific licensed health provider. Using the example of the nursing profession, nurses, as profession-specific licensed health providers, are licensed individually by state law in their respective states. The concept of institutional licensure means that one would be licensed by the organization in which one works. The profession of nursing has been opposed to this concept and successfully fought this phenomena in the early 1970s when there was a national hospital movement to attempt to institutionalize licensure. One critique is that institutional licensure is analogous to the "fox guarding the chicken house." Arguments can be made for strengths and limitations of individual licensure as well as institutional licensure; however, the role of the involved actor determines the responses. If the actor is the health professional, one is threatened by institutional licensure because it detracts from the autonomy of the professional. If the actor is an administrator in a corporate health care organization, one can argue successfully for the merits of institutional licensure (i.e., giving the manager more flexibility in the use of personnel)(*Reforming Healthcare Workforce Regulation* 1995).

In addition to the issue of institutional licensure versus individual licensure, if the health care system moves to having a multiskilled worker versus a profession-specific health professional, the autonomy of each profession is lost. The uniqueness, expertise, skills, socialization, strengths, and other qualities of each profession are lost and melded into a generalist or multiskilled worker caring for patients. This is a concern to nurses because nurses know and respect the difference they make in patients' lives because of their nursing knowledge and expertise. They are aware of the quantitative and qualitative research studies that have demonstrated the positive outcomes of their nursing interventions. While the merits of interdisciplinary work are lauded, the significance of "tribal knowledge" cannot be discounted.

Several organizations have responded to the PEW Commission's report on healthcare workforce regulation. The American Nurses Association (ANA) indicated that while it has applauded many of the Commission's ten recommendations,

> in other areas, the report raises concerns that some of its recommendations or policy options would be an opportunity to weaken regulatory mechanism at a time when the safety and quality of health care services are increasingly threatened by short-sighted attempts by health care institutions and systems to focus primarily on cutting costs and/or increasing profit (Peterson 1997).

The National Council of State Boards of Nursing (NCSBN) has recently moved to initiate a multi-state regulatory process. The purpose is to enhance the individual licensure process for nurses who move state to state, keep licensure status in more than one state, or need to be licensed in more that one state because they are implementing nursing care via telehealth methodologies. This newly initiated state compact system would work like the driver's license model where one state recognizes another state's license. An important part of the impetus behind this initiation of a multistate regulatory process was the intention of being responsive to the types of regulatory concerns identified by the PEW Commission. The NCSBN decision to initiate the interstate compact system indicates an awareness of the need to solve some of the problems raised by the PEW Commission. The increased internationalization of managed care organizations and the expectation of such corporate managers to be able to quickly transfer health personnel from state to state and country to country has been noted by policy makers at the NCSBN. It has been suggested that if regulatory bodies do not initiate change themselves, state regulatory legislation will be introduced and implemented that are economically friendly to corporate organizations and to the global economy.

The ANA is especially concerned about institutional control of nursing practice, the recommended consolidation of medical and nursing boards, and the "competency-based" approach to licensure and practice. If competency-based approaches are used, the ANA wants the full range of competencies required for professional nursing practice to be recognized. Some nurses have critiqued the concern with "competency creep." These nurses would argue that one's fear of competency creep, which is a health care worker's fear that others will take over their duties, can be mitigated by utilizing positive delegation strategies and viewing the situation as a new opportunity. The opposite argument is that it is one more step toward the demise of a profession as more and more duties are released to another (i.e., the proverbial foot in the door) ("18th Annual MAIN Conference" 1997). The concern with using less prepared workers, such as unlicensed assistive personnel (UAP), was expressed by Maffeo (1998). One study of emergency rooms showed that when the registered nurse staffing level was decreased 8% and replaced by UAP's, there was a 400% increased risk of morbidity and mortality. Maffeo (1998) advocated for nursing leadership to take action for quality patient care. The PEW Commission has questioned the use of health professionals on health regulatory boards, alleging that the situation is analogous to the "the fox guarding the henhouse." However, the ANA strongly supports the continued involvement of health professionals in the regulation of their own practice. The ANA plans to continue monitoring the PEW Commission's and other proposals that will affect the practice of nursing and the safety of nursing care for patients.

An example of a state response to the PEW Commission report is a joint study conducted by the Nebraska Nurses Association and the Nebraska State Board of Nursing from 1996 to 1997. This is one of approximately 20 state studies and was funded by the PEW Commission. The purpose was to examine and evaluate Nebraska's current regulatory system in relation to the PEW regulatory recommendations (Oertwich and Burbach 1998). Other goals included identifying policy logic and regulatory principles that could be applied to a spectrum of nursing care providers and to develop a nursing regulatory system that would have five outcomes. These outcomes are (1) protection of the public from harm, (2) promotion of effective health outcomes, (3) facilitation of consumers' access to the most appropriate health care provider, (4) promotion of a cohesive environment among nursing colleagues, and (5) mobility for nurses between practice environment.

The state of Arizona's response to the PEW Commission is seen in a similarly funded project (personal correspondence, Vicky Burbach January 30, 1998). The Maricopa Health Care Integrated Educational System (HCIES) has developed an across-the-continuum health care provider model and has identified competencies of health professionals. The need to reorganize the education of health professionals was recognized after its development. They will be progressing with necessary legislation to change health care regulations to meet concerns raised by the PEW report and to implement the new model (Nelson 1997). Thus, the impact of policy recommendations by the PEW Commission is seen over time in one state. The 1995 study was followed by a state-implemented self-study, which is now being followed up by prepared state legislative regulatory change. Such momentum as this raises concerns for the professional organizations cited in this paper.

National nursing leader Fagin (1998) has noted that nursing groups' negative reactions to the recommendations for change in the regulatory area were caused by concerns about the loss of professional control, the threat of institutional licensure, and the possibility that lesser prepared workers would perform nursing roles. She agreed that these were all legitimate concerns. The irony, however, is the presence of a double-edged sword: for most of the nursing profession, the proposed regulatory changes pose a threat, but for the subset of nurse practitioners, regulatory change could be an asset if the third recommendation, which is the removal of barriers to the full use of competent health professionals, was implemented.

The Coalition on Nursing Futures' Conference held in May 1997 offered another response. The impetus for this Conference was partly because of the release of the PEW recommendations. The following nursing organizations were represented at this Conference: American Academy of Nursing, American

Nurses Association, American Organization of Nurse Executives, Commission on Graduates of Foreign Nursing Schools, National League for Nursing, American Association of Colleges of Nursing, National Federation of Licensed Practical Nurses, National Association for Associate Degree Nurses, state nurses' associations, state boards of Nursing, and nursing organizations representing speciality practice. The two main areas of consensus in response to the challenges raised in this paper are that nurses are responsible for the regulation of nursing practice and that the "driver's license" model of individual licensure should be promoted. Thus, keeping individual licensure versus institutional licensure was promoted as a value.

Another group that has reacted to the PEW Commission's potential threat for institutional licensure is the Nebraska Credentialing Reform Committee Year 2000. The committee was created to purposely study the issues for the licensing and credentialing of all health professionals in Nebraska as we enter the 21st century. The first draft report of that committee has been released. One recommendation was not to endorse the above PEW recommendation of institutional licensure.

Questions about the changing environment and workforce needs are raised by Hurley (1997). He notes that, in the short term, concerted efforts to repeal the layers of professional licensure and certification of health care providers can be observed and he suggests that public policy makers could promote complete deregulation. Deregulation could have adverse effects on certain, if not all, groups of nurses. Certain professions may support some legislative action because they are clinically and economically restricted by present barriers. Advanced practice nurses have worked for many years to decrease barriers to practice. Two groups of advanced practice nurses, nurse practitioners (NPs) and certified nurse midwives (CNMs), will be discussed.

There is no doubt that lack of uniform regulation among states and strict guidelines for reimbursement, prescriptive authority, and physician supervision can be harmful both to the public that demands access to cost-effective primary health care and to the practitioners who wish to provide it (Wilken 1993). NPs and CNMs have been changing state regulatory barriers to practice which have been inherent in many state licensures and scope-of-practice laws and regulations. Scope-of-practice legislation has been attacked as unduly restrictive for many professions and an obstacle to accessing care according to the NCSBN. Nurse practitioners have been battling restrictions in three areas: direct, third-party reimbursement; prescriptive authority; and physician supervision.

Nurse practitioners have demonstrated their ability to provide for 80% of American's primary health care needs at a cost lower than that of physicians' care and without sacrificing quality (Wilken 1993). The ability of NPs to practice independently or work on a collegial footing with physicians is dependent upon reimbursement. The decision to reimburse these two groups rests on three criteria: what services, at what level, and whether third-party payments are made directly to the provider or through a physician. Reimbursement depends in large part on state statutes. In 1993, 26 states had passed mandatory benefit laws requiring private insurers to reimburse NPs directly. In January 1998, all nurse practitioners were able to bill for Medicare Part B directly (Pearson 1998).

Regulatory boards make policy decisions about the power to prescribe for NPs. These policy decisions include determining which providers will be authorized to prescribe drugs and devices, what will be the extent of the authority conferred, and which agency will regulate it. As of January 1997, there were 19 states where NPs could prescribe (including controlled substances) independently of any required physician involvement in prescription writing. Eighteen states require some degree of physician involvement or delegation of prescription writing, including that of controlled substances. In 13 states, NPs could prescribe, with the exclusion of controlled substances, with some degree of physician involvement or delegation or prescription writing. In Illinois, NPs had no statutory prescribing authority (Pearson 1998).

The requirement of physician supervision has a negative impact on the distribution of NPs in rural areas. The legal relationship of physician supervision over NPs varies from state to state. Federal standards for reimbursement to rural health clinics require the presence of a medical doctor once every two weeks in clinics staffed with NPs. However, some statutes have more specific requirements for on-site physician supervision. Certain exceptions may be made for "medically underserved areas" where physicians will not or do not practice but where NPs have stepped in to fill the void. These exemptions make the statutes hypocritical and fundamentally indefensible because exempting supervision in certain areas sends a message that the competence of NPs is determined by where they practice (Wilken 1993).

Policy and structure are key determinants of the changes noted between the professions. The regulation of professional knowledge and practice and the maintenance and perpetuation of institutional boundaries as attributes of professional identity are changing between the professions of nursing and medicine. Over the past 30 years, nursing has developed additional advanced practice models. Collectively known as advanced nurse practitioners, the group consists of certified nurse anesthetists (CRNAs), certified nurse midwives, clinical nurse specialists (CNSs), and nurse practitioners. Each advanced practice group is licensed specifically for their scope of practice. The specific licensure and practice standards allows for nursing autonomy. For example, nurse practitioners are now able to contract directly for health care services in settings such as colleges, businesses, and government agencies (Buppert 1998).

Hurley (1997) sees that such developments as the above will lead to the demise of some health professions and to many health professional schools and programs. He further states the concern noted in the first part of this article—namely, that the preparation of individuals may return to an organization-based training program and that the basic concern is to have a functionally competent individual who is valuable to the organization regardless of the person's education, licensure, and credentials. Such a person is institutionally licensed versus individually licensed and may not have professional independence nor mobility, leading to the potential demise of the autonomy and integrity of a health profession. Changes in the health system have led to HMO's employing physicians. This has resulted in increased economic insecurity for physicians because of downsizing and/or diminished salaries and a resultant decrease in the autonomy of medicine (Robinson 1995; Simon and Born 1996). Hurley (1997, p. 686) recommends that health policy research focuses on the question "How is the landscape of health professional regulation changing, and to what extent might organization licensure replace individual professional licensure?" He also urges research on the aforementioned issue of what degree organizational-based training programs will replace traditional educational institutions in the preparation of health professionals.

Another indicator of the depth of concern about the regulatory recommendations of the PEW Commission can be seen by the response from the Interprofessional Workgroup on Health Professions Regulations (IWHPR): "In an unprecedented consensus, 17 organizations connected with regulation of health care professionals have developed formal recommendations for increasing regulatory effectiveness while protecting the public health" (Woodward 1997, p.1). The Interprofessional Workgroup includes the professional organizations and/or the licensure boards of a wide variety of health care providers—physician assistants, dental assistants, social workers, respiratory therapists, dieticians, occupational therapists, physical therapists, clinical pathologists, speech therapists, chiropractors, health administrators, optometrists, pharmacists, and nurses. These 17 associations represent five million health care providers. The IWHPR alleges that several of the PEW recommendations are undeveloped, lack support data, and would substantially change our present system of state regulation of health professionals but without assuring public protection or cost effectiveness. The IWHPR has three recommendations:

1. Boards with professional and public members, selected for their expertise and commitment and supported by training and adequate resources, make regulations, policies, and individual decisions that assure the public of safe, competent health care providers.
2. Professionals with a solid base of knowledge and skills, gained through academic and clinical education, practice within their corresponding scopes of practice, which may overlap in certain areas with other professions.
3. Competence of health care providers is assured at the entry level through national standardized testing and evidence of both formal professional education and clinical experience (Woodward 1997, p. 2).

These recommendations are intended to improve the process of regulating health professions, increase efficiency, and, most importantly, provide for public protection. The IWHPR frequently alludes to this latter factor as their major goal. However, they also note the significance of this consensus statement because they perceive that their Workgroup represents most of the health professionals who would be affected by the PEW Commission. The response by the IWHPR maintains each state's rights to regulate their own health professionals while facilitating collaboration between regulatory agencies and national certifying agencies. The Workgroup has contacted all states' governors with their response document and has made their professional expertise available to the governors if/and when legislation is introduced in legislatures to change the regulation of health providers. Given the arguments of the ANA described above, the following statements written by the IWHPA regarding scope of practice are significant: "Professional practice is more than the sum of isolated tasks. It requires a defined knowledge base from which a clinician makes critical decisions on why, when, and how to perform or not perform certain tasks" (Woodward 1997, p. 4). Thus, this reasoning would mitigate against the demise of individual professions and the replacement by a multiskilled worker. In their discussion of scope of practice, the IWHPR notes that "a profession's scope of practice must be supported by formal academic and clinical education at the professional level as well as demonstrated competence by examination" (Woodward 1997, p. 4). In summary, this group agrees with many of the PEW Commission recommendations but also has concern with others. The significance lies in the group response of 17 organizations to the PEW document. These organizations expressed concerns reflecting those identified by others discussed in this paper including the NCSBN, ANA, and the work of Hurley.

Another perspective to implementation of the PEW recommendations is that no recommendation should be implemented unless it is evaluated scientifically for its safety prior to implementation in the health care arena. Conceptually, this perspective has been raised about other changes in the health care system. For example, at conferences of the ANA the past few years, nurses have raised this perspective to their concern of decreased staffing ratios in health care settings. They note that new medications can not be used without approval from the Food and Drug Administration. They contend that major changes in the delivery of health care should not be done without similar levels of scientific research. This same conceptual perspective has been raised recently about the practice of gatekeeping in the managed care paradigm, that is, "like any new medically related drug or device, the practice of gatekeeping should be evaluated scientifically" (Personal correspondence, Sharon Hiebert, Florence Listserve Group[1], December 26, 1997). The conceptual argument can be made that PEW recommendations, if implemented, should not simply be implemented by state legislative process; rather, they should be researched empirically for outcomes before mass implementation.

Against the above data of how health professionals regulatory processes are being affected, a recent survey and focus group research study by the American

Hospital Association depicts how consumers evaluate the U.S. health care system (Davidson 1997). Major findings included the following: deep concern about changes that are occurring, decreased access to care, higher costs, lower quality, impersonal care, increasing focus on the "bottom line," and, of specific pertinence to this paper, concern about health providers losing control of their respective scopes of practice and the competence of health care providers. In regard to this latter factor, the competence of health care providers, the study participants held "a strong belief that skilled nurses are being systematically replaced by poorly trained and poorly paid aides....People believe that the profit motive is behind the reduction in nursing care. They are angry at the reversal in health care priorities that they believe this represents" (Davidson 1997, p. 8).

While the focus of this paper has been proposed regulatory changes that will impact the autonomy and integrity of health professions themselves, another aspect is the great increase in other kinds of regulation that have occurred in the health care system in the last three years. Health care providers are subject to many kinds of regulatory law because of the managed care paradigm. Following the failure of national health care reform, "the extensive regulatory activity in 1995-1996 was a watershed, signaling a broader role for state governments in responding to the rapid changes under way in the health care delivery system" (Miller 1997, p. 1102). The major legislative issues that have been regulated at the state level include the following: anti-gag rules, limits on financial incentives, continuity of care, and provider due process. In the first six months of 1996, there were 400 regulatory bills introduced into state legislatures. Of importance to the topic of this paper is that this "regulatory activity is significant, both in terms of its potential impact on clinical practice and its implications for the role of state governments in a rapidly evolving system of managed care" (Miller 1997, p. 1107). Because states have, atypically, responded so quickly on such similar issues, it raises a concern for those agencies who have concerns with the PEW recommendations. Could state legislatures respond as quickly to changing regulations for health professionals that would diminish their autonomy and integrity?

Some who promote the PEW recommendations contend that expanded boundaries will facilitate some patient care and will specifically help the subset of nurse practitioners. While this is positive for the large cohort of nurses, there is a fear of loss of autonomy for the nursing profession and a fear of loss of the uniqueness of nursing's role—especially, if there are outcomes of institutional licensure and a multiskilled worker (Parkman 1997). The final report of the PEW Commission did not contain the institutional licensure recommendation. However, the emergence of these phenomena in both the 1970s and 1990s, and the articles being written about these phenomena in other disciplines, demonstrates the need for health professionals to be aware of this in the changing health care environment.

From this study of one profession, seven action steps that individuals can take if they wish to maintain the autonomy of their profession are proposed:

1. Read widely and be knowledgeable of policies being promoted by policy-makers such as the PEW Commission.
2. Be a member of one's professional association and know their policies and responses to studies, and be active in furthering such policies.
3. Be politically active and respond to proposed regulatory changes in one's individual state concerning licensure and credentialing of one's profession.
4. Participate in state committees that are studying regulation and licensure of health professionals.
5. Be an active political participant if and when changes in license regulation is proposed in one's state legislature.
6. Continue practicing as a competent professional in one's daily work practice and promote the autonomy and integrity of the profession in that daily manner.
7. Conduct research on the effect of regulatory changes on the autonomy of one's profession.

In summary, the rapidly changing health care system is creating challenges to the autonomy and integrity of the individualized health professions. This paper has discussed how it would affect the profession of nursing. The challenges facing one sector in our society—health care—can be instructive to other sectors and professions within those other sectors. Change is pervasive. All professionals need to be aware of how change is affecting society and their role and position in that society.

NOTES

1. The Florence Listserve Group is a group of nurses who implemented a May Day 1998 National Demonstration because of their concern about changes in the health care delivery system and increasing nonsafety for patients.

REFERENCES

"18th Annual MAIN Conference." 1997. *MAINlines* 18(4): 1-22.
Buppert, C. 1998. "Reimbursement for Nurse Practitioner Services." *Nurse-Practitioner: American Journal of Primary Health Care* 23(1): 67, 70, 72-74.
"Critical Challenges: Revitalizing the Health Professions for the Twenty-First Century. 1995. Pp. 1-13 in *The Third Report of the PEW Health Professions Commission*. The Pew Health Professions Commission. San Francisco: University of California.
Davidson, D. 1996. "Public Perceptions of Health Care and Hospitals—A Confidential Report from Dick Davidson to AHA Member CEOs." Chicago: The American Hospital Association.
Fagin, C. 1998. "How Nursing Should Respond to the Third Report of the PEW Health Professions Commission." http://www.ana.org/ojin/tpc5/tpc5_2.html.
Hurley, R. E. 1997. "Managed Care Research: Moving Beyond Incremental Thinking." *HSR: Health Services Research* 32(5): 679-690.

Keddy, B., M. Gillis, P. Jacobs, H. Burton, and M. Rogers. 1986. "The Doctor–Nurse Relationship: An Historical Perspective." *Journal of Advanced Nursing* 11: 745-753.

Maffeo, R. 1998. "Divorcing the UAP: Heresy or Solution?" *The American Nurse* 30 (1):5.

Miller, T.E. 1997. "Managed Care Regulation." *Journal of American Medical Association* 278(13): 1102-1109.

Nelson, N. 1997. "Implications of a Health Care Integrated Educational System for Nursing and Allied Health Students." National Council of State Boards of Nursing Annual Meeting.

"Nursing's Values Challenged by Managed Care." 1998. *Nursing Trends & Issues* 3(1): 1-8.

Oertwich, A. and V. Burbach. 1998. "Nebraska Nurses Regulatory Reform Proposal." Nebraska Nurses Association and Nebraska State Board of Nursing. Lincoln, Nebraska.

Parkman, C.A. 1997. "Health Care Workforce Regulation Reform." *Nursing Management* 28(9): 34-38.

Pearson, L.J. 1998. "Annual Update of How Each State Stands on Legislative Issues Affecting Advanced Nursing Practice." *Nurse-Practitioner: American Journal of Primary Health Care* 23(1): 14-66.

Peterson, C. 1997. "Quality and Staffing Issues Dominate the ANA Agenda." *American Journal of Nursing* 97(2): 53-54.

"PEW Health Professions Commission Revived." 1997. In *Newsletter of the National Council of State Boards of Nursing* Vol. 17, Issue 10, p. 1.

Reforming Health Care Workforce Regulation. 1995. Pew Health Commission. San Francisco: University of California.

Robinson, J. 1995. "The Growth of Medical Groups Paid Through Capitation in California." *New England Journal of Medicine* 333(25): 1684-1687.

Silver, G. 1997. "Editorial: The Road from Managed Care." *American Journal of Public Health* 87(1): 8-9.

Simon, C. and P. Born. 1996. "Physician Earnings in a Changing Managed Care Environment." *Health Affairs* 15(3): 124-133.

Stein, L. and T. Howell. 1990. "The Doctor–Nurse Game Revisited." *New England Journal of Medicine* 322(8): 546-549.

Sweet, D. and I. Norman. 1995. "The Nurse–Doctor Relationship: A Selective Literature Review." *Journal of Advanced Nursing* 22: 165-170.

Wilken, M. 1993. "State Regulatory Board, Regulations, and Midlevel Practitioners in Rural America." PhD dissertation, Department of Political Science, University of Nebraska-Lincoln, Lincoln, NE.

Woodward, S. 1997. "Interprofessional Group Makes Recommendations for Health Care Regulatory Effectiveness." http://www.ncsbn.org/iwhpr/ipwnr197.html.

"Yet Another Survey." 1998. *The American Nurse* 30(1): 8.

CONTROLLING MEDICAL SPECIALISTS:

HOSPITAL REFORMS IN THE NETHERLANDS

Michael I. Harrison and Harm Lieverdink

ABSTRACT

Despite waves of budget cuts and structural reorganizations, physicians in most of Europe have not been subject to managerial controls like those that have emerged in the United States and to a lesser degree in Britain. This paper examines one of the most dramatic instances in which physicians resisted managerial and governmental control and explains recent changes in this pattern. From 1982 through 1992, Dutch hospital specialists successfully blocked efforts by governmental regulators, insurers, and hospital managers to control the specialists' activities so as to reduce hospital costs, integrate care activities, and assure quality. Gradually the specialists began to lose their budgetary and operational autonomy and signed agreements with hospitals and insurers to integrate the specialists' fees into hospital budgets. These new fiscal arrangements increase the hospital managers' economic control over specialists and create possibilities for the enhancement of other types of managerial control. The organizational and political forces behind these changes are analyzed, along with their implications for future patterns of professional control in the Netherlands and for the integration of professional and managerial functions within Dutch hospitals.

Research in the Sociology of Health Care, Volume 17, pages 63-79.
Copyright © 2000 by JAI Press Inc.
ISBN: 0-7623-0644-0

INTRODUCTION

Sweeping changes in health care organization throughout the West directly threaten the professional prerogatives of many physicians and increase the possibilities for control by governmental regulators, hospital managers, and insurers (Hafferty and McKinley, 1993). Despite this general trend, physicians in most European countries have remained free of managerial controls like those that have emerged in the United States (Bjorkman 1989) and to a lesser degree in Britain (Calnan and Williams 1995; Glendinning, Chew, and Wilkin 1994; Harrison and Pollitt 1994). Moreover, in both Europe and North America relations between the health professions and those occupational groups who seek to manage and regulate them are more dynamic and varied than is sometimes thought. These relations depend on labor markets (Bacharach, Bamberger, and Connely 1990) and on interactions, coalitions, and power shifts among actors shaping health policy (Dohler 1991; Kenis and Schneider 1991; Light 1996; Schwartz and Busse 1997; van de Ven 1997). Struggles and negotiations within and between professions also decisively shape relations among professions, managers, and other key actors (Abbott 1988; Harrison 1994). As a result of these internal and external developments, some groups of physicians have successfully resisted the imposition of nonprofessional control over their activities (e.g., Barnett, Barnett, and Kearns 1998; Harrison 1993, 1999).

To illuminate the forces affecting relations between physicians and nonprofessional managers and regulators, we examine struggles by Dutch hospital physicians to maintain budgetary and operational autonomy in the face of growing managerial constraints. These constraints stemmed from moves by governmental regulators, insurers, and hospital managers to curtail rising hospital costs, integrate care activities, and assure the quality of care. We focus mainly on the fifteen-year struggle over governmental regulation of the specialists' fees and on the forces that are producing fundamental changes in the specialists' occupational and budgetary status. From 1982 through 1992 Dutch hospital specialists successfully blocked efforts by governmental regulators, insurers, and hospital managers to control the specialists' activities. Gradually the specialists began to lose their budgetary and operational autonomy and signed agreements with hospitals and insurers to integrate the specialists' fees into hospital budgets. These new fiscal arrangements increase the hospital managers' economic control over specialists and create possibilities for the enhancement of other types of managerial control.

Three questions guide the analysis of these developments:

1. What factors account for the ability of Dutch hospital specialists to resist governmental regulation and managerial control for such a long period?
2. What accounts for the shift in the last few years toward greater integration of specialists' fees within hospital budgets and toward growing involvement of specialists in decisions relating to hospital management?

3. Are these changes in budgeting and physician involvement intensifying and extending managerial control over the specialists and undermining professional control?

Question one is addressed in the first part of the analysis presented below. That part of the paper examines forces that made it possible for hospital physicians to resist governmental and managerial control until a few years ago. The second part of the analysis, which responds to the second question, accounts for recent local and regional changes in the budgeting and managing of physicians' services. These developments are explained in terms of changes in the power and tactics of governmental actors and of the specialists; changing interactions among hospital managers, insurers, and physicians; and the emergence of new cognitive schemes among policy actors. The concluding part of the paper takes up the third question. It considers whether recent developments in and around the hospitals are leading to intensified managerial control over hospital physicians.

The analysis that follows draws on several data sources. Among them were archives of the main national organizations engaged in the struggle over the specialists' fees—including the Ministry of Health, major health insurers, the National Association of Medical Specialists (LSV in Dutch), and the quasigovernmental, Central Agency for Health Care Tariffs. Conclusions drawn from archival sources for 1982 through 1992 were validated through open-ended interviews with individuals who had represented these national organizations during negotiations over physicians' earnings (Lieverdink 1999, Lieverdink and Maarse 1995). In addition, in 1994 (Harrison 1995) and 1996, Harrison conducted semistructured interviews with members of the LSV, the Ministry of Health, the National Hospital Association, and with health services researchers. He also interviewed hospital administrators and specialists in three hospitals that were in the process of introducing new arrangements for remunerating specialists. Relevant publications and documents were also reviewed.

MAINTENANCE OF PROFESSIONAL CONTROL: 1982–1992

The Struggle over the Fees and Organizational Status of the Specialists

As part of its broad campaign to rein in health care expenditures, the Dutch government began to impose prospective budget ceilings on hospital expenditures during the early 1980s.[1] Despite these efforts, for more than a decade, payments to hospital specialists and associated costs remained largely beyond control of the government and the hospital managers. Let us examine how hospital specialists managed to block reforms of their remuneration system for so long and how they remained aloof of managerial decision making structures in the hospitals.

During the 1980s, nearly all hospital specialists in general hospitals were self-employed and were paid on a fee-for-service basis (Grunwald and Mantel 1992, p. 108). Salaried specialists, most of whom worked in acute hospitals, were concentrated in areas like dermatology, pediatrics, rehabilitation, clinical chemistry, rheumatology, psychiatry, and radiology.

In order to obtain a contract to practice in a hospital, the self-employed specialist purchases membership in a specialty group (*maatschap*), in which he or she works on a full-time basis. This group usually has from three to six members (Saltman and de Roo 1989). These institutional arrangements generate longterm ties to a single hospital for most specialists and result in very low mobility between hospitals.

The specialty group serves as the primary unit within which specialists are organized in general hospitals. It has no formal hierarchy, and members are treated as equal partners headed by a colleague of professional distinction. Within these groups, there are no formal procedures for conflict resolution or grievance processing. Until recently, very few specialists outside of university hospitals held positions in the hospital's formal administrative or professional hierarchies. Each hospital had a medical board composed of physicians, but this board had little power over the medical staff. Instead, most clinicians enjoyed the paradoxical situation of having "power over almost everything" without being administratively responsible for anything (Saltman and de Roo 1989, p. 785).

Until recently, the earnings of the majority of hospital specialists, who work on a fee-for-service basis, were not included in hospital budgets. Instead specialists' fees were based on nationally agreed-upon fee schedules, and were billed directly to the insurers (sickness funds and private insurers), which reimbursed physicians for their work. Prior to 1982 the specialists' fees were negotiated between the representative associations of the specialists, particularly the National Association of Medical Specialists (LSV) and the associations of sickness funds. The results of these negotiations were subject to approval by the Sickness Fund Council, which was made up of representatives of the parties interested in the health system. The specialists and the sickness funds dominated the negotiation process and typically shifted the costs of fee increases onto the employers and the public.

Throughout the late 1970s and early 1980s, the Secretary of Health and the media denounced the excessive incomes of specialists. Finally, the government took action against the specialists' fees through The Health Care Tariffs Act of 1982. This bill unified procedures for negotiating specialists' fees and shifted negotiations into the Central Agency for Health Care Tariffs, a quasigovernmental setting in which the government had more say over the specialists' fees and earnings.

Fearing a backlash against the specialists, the LSV accepted the new arrangements for setting fees and entered into lengthy negotiations with the government in search of a formula that would hold down the total earnings of the specialists to a level deemed reasonable by the concerned parties.[2] In 1982, the LSV and the

Ministry of Health agreed that regressive fees—which declined as practice volume increased—could be imposed on practitioners in certain highly paid specialties.

In practice the specialists refused to cooperate with the reporting mechanisms needed to administer the regressive fee system. When the hospital physicians refused to implement the agreements, the government and the insurers proposed incorporating physicians' charges within hospital budgets and thereby imposing budget caps on specialists' earnings. The LSV denounced this proposal as a threat to their members' income and autonomy (Lieverdink and Maarse 1995). The proposal was also opposed by the national hospital association and was eventually defeated. The government then imposed direct cuts in physicians' earnings by reducing fixed payments for office costs while leaving the fee-for-service system in tact.

In response to actual and proposed inroads into their incomes, the specialists began to make unprecedented use of strikes and sanctions (Kirkman-Liff 1989). In 1986 the specialists undertook the first specialists' strike in the history of the Dutch medical profession (de Roo 1988), which was followed by additional wild-cat strikes (Glaser 1991). Moreover, the specialists continued to refuse to provide the data needed to calculate the amounts they were supposed to refund when budget caps were exceeded. As an additional sanction during these conflicts, many specialists canceled their contracts with the sickness funds and began charging patients directly.

By 1988, earnings-regulation agreements between the government and the specialists had collapsed. Partly as a result of frustration over the inadequacies of top-down regulation, government health policy was moving toward the introduction of managed competition as a way of controlling health costs (de Roo 1995; Harrison 1995; van de Ven 1997). In response to the uncertainties and threats posed by this new policy thrust, the national representatives of specialists, insurers, and hospitals briefly reverted to a cooperative stance. Their negotiations yielded the Five Parties Agreement, which abandoned regressive fees and fee cuts for specialists. In their place, the specialists accepted the principle of a cap on their earnings—to be obtained through a national cap on the volumes of specialists' activities. In addition, the earnings of the most highly paid specialties were to be reduced while those of underpaid groups were to be boosted, so as to reduce inequalities between specialties.

Although ostensibly in place until 1992, the Five Parties Agreement was never accepted by large parts of the medical community. Rather than restricting their activities, specialists continued to increase their volume of activity and apparently used creative bookkeeping to increase the number of *reported* procedures. Moreover, the Five Parties Agreement led to schisms within the specialists' organizations. Highly paid, elite specialists, many of whom leaned to the political right, withdrew from the LSV in protest over its failure to preserve their earnings and represent their interests. Another group of left-leaning, self-employed specialists attacked the LSV for paying too much attention to revenues and neglecting quality of care. Salaried physicians and junior doctors became increasingly vocal through yet a third specialists' organization, which sought to represent their special interests.

Analysis

For eleven years, the specialists succeeded in blocking meaningful reform of their terms of remuneration and employment, resisted cuts in their earnings, and remained largely aloof from involvement in hospital management. The specialists' capacity to resist both governmental initiatives and public opinion derived from several sources. First, the specialists alone had the knowledge and authority to determine what treatments and diagnostic procedures their patients needed. Hence, only the specialists could legitimately determine the total volume of their activities. No countervailing authority to review and question these decisions existed in the Ministry of Health, the insurance companies, or the hospitals.

Second, the hospital physicians effectively mobilized both nationally and locally to subvert and sometimes directly resist attempts to implement regressive fee schedules and make direct cuts in their earnings. Third, the specialists received support in their struggle from powerful policy actors—including the national hospital association. Individual hospital directors also usually sought to avoid confrontations with their specialists and favored increasing hospital production so as to meet patient demand. In contrast, the insurers usually supported government efforts to contain hospital costs and curtail specialists' earnings. However, the insurers lacked power over the hospitals and avoided aggressive moves to curtail the specialists earnings, because patients as well as physicians would have opposed such moves.

INTEGRATION OF PHYSICIANS INTO HOSPITAL BUDGETING: 1992–1998

Major Changes in the Specialists' Status

The specialists' wall of opposition to occupational, organizational, and budgetary change began to crumble in the early 1990s. We will first describe the most important recent changes affecting the specialists' status. Then we will explain these developments in terms of changes in the power positions, interactions, and cognitive schemes of the specialists and other major policy actors.

As the Five Parties Agreement foundered, the Secretary for Health and other governmental actors abandoned their efforts to achieve fiscal constraint over the physicians' earnings through professional self-regulation. Instead, in 1992, the government introduced changes in the Health Tariffs Act that allowed it to cut physicians' fees unilaterally when costs for physicians' services exceeded budgeted levels.

Backed by this legislative change, the Ministry of Health mandated budget cuts for the specialists, like those that had previously succeeded in holding down hospital expenditures. In April 1993, Secretary for Health Simons cut the fees for most specialists by 12% and introduced even deeper cuts for some highly paid specialists. This move set off a storm of sanctions, legal actions, and protests—none of which succeeded in blocking the decree. In the end, the Ministry of Health succeeded in imposing fiscal restraint on the specialists but produced considerable discontent within the medical profession (Lieverdink and Maarse 1997).

During the early 1990s, the government also formalized a longstanding view that the specialists' earnings should be treated as an integral part of hospital budgets, rather than being billed and reimbursed separately. This view was expressed most clearly in 1993 by an advisory committee headed by the former prime minister, Biesheuvel. That committee recommended that physicians be paid on a salarylike basis, rather than retaining their status as independent practitioners, who are paid fee-for-service. In 1994, the Biesheuvel Committee's recommendations were approved by the Ministry of Health and by parliament.

Despite opposition from the LSV, these recommendations were recently introduced into a proposed law. In January 1998, a weak version of the committee's view was adopted by one of Holland's two house's of parliament as an amendment to the pending Sickness Fund Act.[3] Although the amendment did not address the physicians' occupational status in the hospitals, it did give hospital management overall responsibility for hospital finance. Moreover, by treating the hospital as an single, integral organization, the bill formally marks the end of the era of separate budgeting for physicians' fees and for other hospital costs.

The most decisive development affecting the physicians' occupational status was the drive by the Ministry of Health to encourage local or regional agreements covering specialists' fees and the volume of their activities. Beginning in 1994, Mrs. Borst-Eiler, the newly appointed Minister of Health, threatened new rounds of fee cuts unless the specialists entered into contractual agreements with their hospital managers and with insurers. In exchange for creation of these new "triangular" (i.e., three-way) agreements, the Ministry of Health offered to maintain the current level of specialists' earnings for a three-year period.

Faced with the threat of more earnings cuts and the prospect that the government might turn self-employed specialists into salaried employees, as was recommended by the Biesheuval Committee, the specialists quickly began to cooperate with proposals for local agreements. By 1997, nearly all general hospitals had developed agreements with their physicians, and the majority of self-employed specialists (i.e., members of *maatschapen*) in general hospitals had entered into local or regional budgetary agreements (Kahn 1997).

The new agreements created fixed budgets for specialist care. Using the 1994 fee schedule, the parties specified volumes of admissions, outpatient treatments, and sometimes other variables, so as to produce a fixed prospective budget for each hospital's expenditures for specialty care. Some of the three-way agreements introduced provisions for reducing earnings differentials between specialties, but most left these differentials in tact (National Hospital Institute 1996). In addition, as per governmental requirement, the new agreements included statements about steps toward quality improvement. In practice, the governmental regulators did not try to evaluate the scope or implementation of quality projects (National Hospital Institute 1996). Moreover, hospitals were able to fulfill their contractual obligations by committing themselves to continue ongoing projects as well as introducing new ones.

In parallel with these important developments, administrative changes took place in some hospitals that also led to greater integration of hospital specialists into managerial communication, decision making, and budgeting. During the early 1990s some hospitals experimented with involving physicians in managerial decision making. In most cases these experiments focused on decisions concerning hospital policies, rather than on day-to-day administration (Versluis 1993). Another influential development occurred in some large hospitals, which adopted divisional structures. The few physicians who accepted positions as heads of divisions and of other hospital functions, took on more line management responsibilities, along with their responsibilities for overseeing medical care within their division (Pool 1992). In like manner, physicians responsible for coordination between their specialty group and other specialty groups or units and physicians who took part in negotiations between hospitals and insurers became more involved in managerial decision making and more aware of budgetary and administrative constraints on medical work.

Analysis

To account for the changing status of hospital physicians, we must consider shifts in the power positions and tactics of the major national actors in the struggle over the physicians' fees and status; changing patterns of interaction among local actors; and the gradual adoption by actors of new concepts and cognitive schemes.

Perhaps the single most powerful explanation for the sudden turnabout in the status of the hospital physicians lies in the fragmentation of their national organization and its associated loss of political power. When the Five Parties Agreement of 1989 failed to satisfy important constituencies within the LSV, that organization split into quarrelling factions. These factions backed different ideologies and took divergent positions on the earnings inequalities that prevailed among the specialists. Several years later, the representatives of these factions patched over their differences, but the reconstituted LSV was a mere shadow of the former organization. Its membership had dropped radically. Furthermore, its coalitional structure reduced its tactical options, and its leadership encountered difficulties in rallying the rank and file to present a unified front against governmental initiatives.

The LSV also suffered from a loss of prestige due to the militant sanctions which it promoted and due to the ideological and tactical inflexibility of its spokespersons. A further blow to the power of the specialists at the national level came as physicians within individual hospitals began organizing their own unions in response to local budgetary initiatives. All of these developments can be understood as part of the erosion of neocorporatist structures in Dutch health politics, through which national representatives of interest groups once reached agreements that determined the policy options for their members (van der Grinten 1996).

The specialists' loss of influence also stemmed from a deepening conflict of interest between them and hospital managers, who had traditionally supported the specialists' budgetary autonomy and their struggle to increase their earnings. In response to budget ceilings, hospital managers imposed increasing constraints on the volume of hospitalization and on the personnel, funds, and facilities available to support medical work. At the same time, the continuation of fee-for-service payments to specialists and the use of regressive fee schemes created strong incentives for physicians to increase hospital admissions and expand the volume and range of treatments and diagnoses. Seeking to resolve this fundamental conflict of interest, the National Hospital Association expressed support for the recommendations of the Biesheuval Committee, thus ending the Hospital Association's tradition of supporting specialists' attempts to preserve their status as "free entrepreneurs" working within the hospitals.

As the specialists' power waned at the national and local levels, the Ministry of Health gained power within the government and developed new tactics for influencing the specialists. Already in the late 1980s, policy development within the Ministry of Health and the government began to bypass the LSV, rather than soliciting the specialists' inputs throughout the policy formation process as it had done in the past. Thus the government developed its policies on health budgeting and its positions on regulated competition in forums in which the specialists and other major stakeholders in the health system were not represented.

The influence of the Ministry of Health over the specialists increased markedly after the elections of 1994. The newly appointed Minister of Health, Welfare and Sport, Mrs. Borst-Eilers, took charge of the health portfolio herself, rather than delegating it to a Secretary for Health, as had been done under the two previous governments. This change meant that a minister who was knowledgeable and personally concerned with health affairs took part in all cabinet meetings, rather than simply being asked to participate when health issues were discussed. In this connection, it is noteworthy that Mrs. Borst-Eilers is a physician with personal ties to prominent members of the medical community and considerable experience in medical administration.

A capable tactician, Mrs. Borst-Eilers and her staff reverted to an incremental style of guiding health politics that harked back to less conflict-ridden periods in national health administration and contrasted with the attempts of health secretaries Dekker and Simons to make a radical overhaul of the system by introducing regulated competition. The new minister's carrot-and-stick tactics effectively steered the physicians toward compliance with local three-way contracts. These tactics thus produced more tangible results with less disruption than had the confrontational tactics used by the previous Secretary for Health.

An additional boost to the Health Ministry's power came from the backing given its policies by the Biesheuval Committee and the Dutch parliament. For years, a majority of parliament members had sought ways to curtail the specialists' earnings. Formal support for the Ministry came in the parliament's endorsement of changes in the Health Tariffs Act and in its endorsement of the Biesheuval Committee. These political developments provided the Ministry with a clear mandate for introducing change and allowed Borst-Eilers and her staff to negotiate with the specialists from a stronger political position than that of the previous sectaries of health.

Interactions among policy actors at the local and regional levels also contributed to changes in the specialists' status. During the late 1980s and early 1990s, the government placed increasing pressure on the publicly funded sickness funds (insurers) to contain costs. The insurers, in turn, began to transmit these pressures to hospital managers, from whom they purchased health care. In a similar fashion, managers in insurance companies and in hospitals grew worried about the probable impacts on their operations of regulated competition among insurers, and to a lesser degree, about competition among providers. More recently, hospital managers began to adjust their operations so as to take part in integrated care programs. In keeping with current governmental thinking, experiments in integrated care seek to reduce hospitalization and to increase cooperation between hospitals wards, outpatient facilities, community-based health care, and social services. Some experiments in integrated care contain possibilities for money to follow the patient's choice of care location.

In response to current and anticipated external developments like these and in reaction to direct cuts in their hospitals' operating budgets, hospital managers tried to enlist physicians in cost-containment efforts and in the restructuring of hospital services. At first physicians were almost universally antagonistic to these overtures. However, by the early 1990s, some specialists had developed a more pragmatic orientation toward their hospital directors and toward management. These specialists began to take part in decisions concerning areas such as staffing, the introduction of cost savings, and treatment priorities. In many hospitals, managers, and specialists started to cooperate in the development and promotion of additional hospital services, as a way of enhancing physician income and revenue flows, and in response to pressure from the insurers and the Ministry of Health to substitute ambulatory care for inpatient care and develop integrated care packages (van der Grinten, Meurs, and Putters 1998). Some hospitals offered more rapid treatment in exchange for higher insurance fees. Some general hospitals also entered the growing field of home care for the aged, mainly by offering nursing services.

The old budgeting arrangement, with its radical separation of hospital and physicians' budgets, created many barriers to restructuring care and to developing new sources of revenue for the hospital and its physicians. In contrast, the new three-way contracts created opportunities for restructuring care. Hence initiatives for restructuring care and the development of local contracts were mutually reinforcing.

The changing relations between physicians, regulators, and managers examined so far reflect changes in the concepts and cognitive schemes used by all parties, as well as in their power positions and in their interactions. During the past decade, health politicians, insurers, managers, and even physicians have been increasingly exposed to cognitive schemas and forms of discourse that draw concepts and frames from the worlds of business, marketing, and industry and apply these schemas to health care.

The first wave of new concepts dealt with cost-efficiency. Initially hospital physicians completely rejected considerations of cost. Most specialists simply carried on their clinical work as usual and paid little attention to the ways that their clinical activities affected hospital costs. Instead, as our interviews show, they stated that worrying about the costs of medical practices contradicted the requirement that the physician should carry out the best treatment possible— whatever the cost.

However, as cost constraints deepened, hospital physicians became increasingly aware of the cost-implications of their clinical decisions. By the early 1990s some physicians began to accept joint responsibility for hospital costs and began to get involved in applying business and budgetary thinking to medical work. Few hospital physicians adopted these managerial schemas without reservations. For most of those who participated in management, involvement was simply preferable to allowing management to impose cuts in staff and resources without consulting with physicians.

A second wave of new concepts and schemas dealt with the marketing of medical services. These concepts were fostered by governmental proposals to create regulated competition among insurers and in some cases among providers. At first, these notions were alien to insurers and hospital managers as well as to physicians. However, insurers and managers soon began to use terms from marketing and to think more entrepreneurially about generating revenues and attracting and retaining clients. Some physicians were also exposed to these approaches, through their collaboration with managers and insurers and through programs aimed at training them in the use of business concepts. In a similar fashion, some physicians and managers began to adopt the vocabulary and perspective of business strategy, either as a result of participation in training programs, exposure to market-oriented governmental policies, or interaction with other business-oriented factors, such as drug companies and medical entrepreneurs.

Yet another stream of new concepts and models deriving from business and industry deals with assessment and assurance of quality. At least in theory, the market-oriented reforms of the late 1980s gave insurers the ability to chose services based on quality considerations. Moreover, during this period, the Dutch medical profession responded to growing governmental interest in quality issues by assuming responsibility for quality assurance activities at both the national and hospital level. Despite the mandate to include quality assurance in the new three-way contracts, quality assurance in the hospitals continues to be dominated by the physicians and remains largely professional and educational in character (Harrison 1999; Lieverdink and Maarse 1997).

Although the extent of change is hard to gauge, few physicians and managers have remained entirely untouched by these new ways of thinking and talking about health services. One study of the five hospitals that pioneered the integration of physicians' fees into hospital budgets found definite changes in physicians' attitudes and behavior between 1974 and 1997 (Sickness Fund Council 1998) Although most physicians continued to express a preference for the fee-for-service system, they showed greater cost awareness. Examination of hospital records showed that the physicians reduced their volume of treatments and readmissions to hospital in order to meet hospital budgetary targets.

The gradual changes that are occurring in discourse and cognition both contribute to change in the physicians' role and bargaining position and reflect other sources of change in the physicians' situation. These shifts in shared discourse and cognition help redefine the prevailing rules of the game, which in turn influence the clinical actions of physicians and shape interactions between physicians and managers.

CONCLUSION

In conclusion, what do the developments reported here suggest about the third issue noted at the outset of this paper: Are the reported changes in budgeting and physician involvement intensifying and extending managerial control over hospital specialists and undermining professional control? Let us consider briefly how recent developments affect the following four types of professional control (Harrison 1993, 1994) at the national level and within local hospitals:

1. *Economic control* refers to the professionals' ownership and control over the physical and technical facilities needed to produce their services. Economic control also refers to the degree to which professionals can set the terms of remuneration for their services and determine the conditions under which the services are delivered (e.g., whom to treat and when).

2. *Strategic control* concerns the ability to set or influence the goals, policies, and objectives of professional work.

3. *Administrative control* deals with ability to manage the organizational units in which professionals work and to coordinate relations between units.

4. *Operational control* entails the freedom to be guided by professional judgement and standards in professional practice as opposed to being subject to supervision and control by managers.

There is no doubt that at the national level, governmental regulators have achieved greater economic control over the total earnings and terms of remuneration of hospital specialists. In like manner, the Ministry of Health strengthened its hand in health policy formation, and the national physicians' organizations lost much of their former capacity to exercise strategic control over health finance and other forms of health policy. Still, the Ministry of Health did not attempt to dictate the degree of administrative control to be exercised by physicians at either the national or the hospital level. Furthermore, despite its encouragement of quality assurance, the government did not challenge the medical profession's virtually unquestioned control over decisions affecting clinical practice and diagnosis and over the assurance of clinical quality.

Emerging patterns of control at the hospital level are more complex. In the economic sphere, the new contracts provide hospital managers and insurers with much more influence over specialists' working conditions, their terms of remuneration, and their total earnings. In the end, the shift toward integrated budgeting may spell the end of the independent specialist groups (*maatschappen*). One indication of this possibility is a growing tendency for members of specialist groups to shift to salaried status.

The growing involvement of physicians in hospital management can affect the administrative and strategic control of hospital specialists in several different ways. On the one hand, participation in management can spell the end of autonomous decision making about staffing, contacts with other units, and other administrative issues. From this standpoint, management involvement serves as a form of cooptation whereby physicians are brought into hospital management forums and thereby become subject to managerial considerations and priorities that they could previously ignore. The diffusion within hospitals of concepts and frames taken from the fields of industry and business could also contribute to such a process of cooptation.

On the other hand, the growing involvement of physicians in hospital management provides new opportunities for physicians to influence administrative and strategic decisions. For example, physicians have become active in helping managers justify charges and costs to insurers. Furthermore, physicians are beginning to take an active role in the development of new hospital services and policies. In a similar fashion, specialists continue to show initiative in the acquisition of new medical technologies and in the adoption of new forms of treatment and diagnosis.

Yet another possibility is that managerial-professional relations will not become a zero-sum game. Instead, as van der Grinten and his colleagues (1998) suggest, as hospitals become more entrepreneurial, managers have "much more to manage." Hence, the total power of the organization will increase without producing a fundamental shift in the division of power between managers and physicians. The realization of this scenario depends on the ability of hospitals to increase their budgets and activities, despite strong external pressures to cut down on hospitalization and shift care to the community.

Hospital managers and insurers thus far have not used the new contracts to exercise leverage over the specialists' clinical practice. As in the past, operational control over clinical practice remains almost exclusively in the hands of physicians. Furthermore, managers and physicians have not introduced dramatic changes in their approach to peer regulation. For instance, physicians alone take part in quality assurance activities and do not systematically use quality data to generate normative pressure on physicians, specialty groups, or entire hospitals. Nor do physicians share quality data with managers or lay persons.

Nonetheless, the new agreements make it virtually impossible for physicians to ignore the cost considerations of their work. Moreover, these and other concepts from business and industry are increasingly shaping discussions about the nature of medical work among physicians and between them and managers.

Eventually Dutch hospital managers and insurers may try to assess medical practices in terms of the protocols for best practice that are being developed by national medical specialty associations or in terms of comparative data on hospital practices and outcomes. Such moves, which have already occurred in the United States (Ovretveit 1996) and have been proposed elsewhere (Berg 1996; Rappolt 1997), would certainly erode the occupational autonomy and operational control of hospital specialists. Another longterm possibility is that the development of integrated forms of care will further weaken the economic position of hospital-based specialists as care is transferred out of the hospitals and as community-based physicians and other types of health practitioners take over more care functions.

ACKNOWLEDGMENTS

This chapter is a revised version of a paper presented at the annual conference of the European Healthcare Management Association, Dublin, Ireland, June 1998. The first author acknowledges support for portions of this research from the Israel National Institute for Health Policy and Health Services Research and Bar-Ilan University. He also owes thanks to Eric Konnen and the staff of the National Hospital Institute in Utrecht for greatly facilitating two rounds of interviews in the Netherlands. The views expressed in the chapter are the sole responsibility of the authors and not of their employers or funding sources.

NOTES

1. Background on Dutch health care and the reforms of the 1980s appear in de Roo (1988), Harrison (1995), Organization for Economic Cooperation and Development (1992), and Schrijvers and Kodner (1997).
2. For further details on the struggle over specialists fees, see Harrison (1995) and Lieverdink and Van der Maade (1997).
3. The amendment has encountered opposition in the Parliament's First Chamber and may undergo further modifications before final passage.

REFERENCES

Abbott, A. 1988. *The System of Professions: An Essay on the Division of Expert Labor.* Chicago: University of Chicago Press.

Bacharach, S., P. Bamberger, and S. Connely. 1990. "Negotiating the 'See-saw' of Managerial Strategy: A Resurrection of the Study of Professionals in Organization Theory." *Research in the Sociology of Organizations* 8: 217-238.

Barnett, J.R., P. Barnett, and R. Kearns. 1998. "Declining Professional Dominance? Trends in the Proletarianization of Primary Care in New Zealand." *Social Science and Medicine* 46: 193-207.

Berg, M. 1996. "Problems and Promises of the Protocol." *Social Science and Medicine* 44: 1081-1088.

Bjorkman, J. 1989. "Politicizing Medicine and Medicalizing Politics: Physician Power in the United States." Pp. 28-73 in *Controlling Professionals: The Comparative Politics of Health Governance,* edited by G. Freddi and J. Bjorkman. New York: Sage.

Calnan, M. and S. Williams. 1995. "Challenges to Professional Autonomy in the United Kingdom? The Perceptions of General Practitioners." *International Journal of Health Services* 25: 219-241.

de Roo, A. 1988. "Netherlands." Pp. 215-228 in *The International Handbook of Health-Care Systems,* edited by R. Saltman. New York: Greenwood Press.

_____. 1995. "Contracting and Solidarity: Market-oriented Changes in Dutch Health Insurance Schemes." Pp. 45-64 in *Implementing Regulated Markets in Health Care,* edited by R. Saltman and C. von Otter. Buckingham: Open University Press.

Dohler, M. 1991. "Policy Networks, Opportunity Structures and Neo-Conservative Reform Strategies in Health Policy." Pp. 235-296 in *Policy Networks: Empirical Evidence and Theoretical Considerations,* edited by B. Marin and R. Mayntz. Boulder, Colorado: Westview Press.

Glaser, W. 1991. *Health Insurance in Practice: International Variations in Financing Benefits, and Problems.* San Francisco: Jossey-Bass.

Glendinning, C., C. Chew, and D. Wilkin. 1994. "Professional Power and Managerial Control: The Case of GP Assessments of the Over-75s." *Social Policy and Administration* 28: 317-332.

Grunwald, C.A. and A.F. Mantel. 1992. "The Netherlands." Pp. 104-115 in *European Health Services Handbook,* edited by N. Leadbeater. London: Institute of Health Services Management.

Hafferty, F. and J. McKinley, Editors. 1993. *The Changing Character of the Medical Profession: An International Perspective.* New York: Oxford University Press.

Harrison, M. 1993. "Medical Dominance or Proletarianization? Evidence from Israel." *Research in the Sociology of Health Care* 18: 73-96.

_____. 1994. "Professional Control as Process: Beyond Structural Theories." *Human Relations* 47: 1201-1230.

_____. 1995. *Implementation of Reform in the Hospital Sector: Physicians and Health System Reforms in Four Countries.* Tel Hashomer, Israel: Israel National Institute for Health Policy and Health Services Research.

_____. 1999. "Health Professionals and the Right to Health Care." Pp. 81-99 in *The Right to Health Care in Several European Nations,* edited by Andre den Exter and Herbert Hermans. London: Kluwer Law International.

Harrison, S. and C. Pollitt. 1994. *Controlling Health Professionals: The Future of Work and Organization in the NHS.* Buckingham: Open University Press.

Kahn, P.S. 1997. "On Triangles and Marriages: Considerations on the Integration of Medical Specialists Care Act." [in Dutch] *Zorg en Verzekering* 5: 561-569.

Kenis, P. and V. Schneider. 1991. "Policy Networks and Policy Analysis: Scrutinizing a New Analytical Toolbox." In *Policy Networks: Empirical Evidence and Theoretical Considerations,* edited by B. Marin and R. Mayntz. Boulder, Colorado: Westview.

Kirkman-Liff, B. 1989. "Cost Containment and Physician Payment Methods in the Netherlands." *Inquiry* 26 (winter): 468-482.

Lieverdink, H. 1999. "Collective Decisions, Interests and Law: Decision Making on Fees for Medical Specialists in the Netherlands" [in Dutch]. Ph.D. Dissertation. Faculty of Health Sciences, Maastricht University, Maastricht, The Netherlands.

Lieverdink, H. and H. Maarse. 1995. "Negotiating Fees for Medical Specialists in the Netherlands." *Health Policy* 31: 81-101.

_____. 1997. "A New Form of Health Care Contracting in the Netherlands: Better Conditions for Health Care Improvement?" Paper presented at the Centre for Health Economics and Policy Analysis (CHEPA) Conference, 21-23 May. Hamilton, Canada.

Lieverdink, H. and J. van der Made. 1997. "The Reform of the Health Insurance Systems in Germany and the Netherlands: Dutch Gold and German Silver." Pp. 109-135 in *Health Policy Reform, National Variations and Globalization,* edited by C. Altenstetter and J.W. Bjorkman. London: Macmillan.

Light, D. 1996. " Countervailing Powers: A Framework for Professions in Transition." Pp. 25-66 in *Health Professions and the State in Europe,* edited by T. Johnson, G. Larkin, and M. Saks. London: Routledge.

National Hospital Institute (NZI). 1996. *Medical Specialists under a Fixed Budget. Evaluating the Effects of Local Initiatives.* [In Dutch]. Utrecht: National Hospital Institute.

Organization for Economic Cooperation and Development (OECD). 1992. *The Reform of Health Care: A Comparative Analysis of Seven OECD Countries.* Health Policy Studies No. 2. Paris.

Ovretveit, J. 1996. "Informed Choice? Health Service Quality and Outcome Information for Patients." *Health Policy* 37: 75-90.

Pool, J. 1992. "Hospital Management: Integrating the Dual Hierarchy?" *The International Journal of Health Planning and Management* 6:193-207.

Rappolt, S. 1997. "Cliinical Guidelines and Fate of Medical Autonomy in Ontario." *Social Science and Medicine* 44:977-987.

Saltman, R. and A. de Roo. 1989. "Hospital Policy in the Netherlands: The Parameters of Structural Stalemate." *Journal of Health Politics, Policy and Law* 14(4, winter): 773-795.

Schrijvers, A.J.P. and L. Kodner, Editors. 1997. *Health and Health Care in the Netherlands.* Utrecht: De Tijdstroom.

Schwartz, F. and R. Busse. 1997. "Germany." Pp. 104-118 in *Fixing Health Budgets: Experience from Europe and North America,* edited by F. Schwartz, H. Gelnnerster, R. Saltman, and R. Busse. New York: Wiley.

Sickness Fund Council. 1998. *Evaluation of Experiments in the Payment of Medical Specialists* [in Dutch]. Amstelveen: Sickness Fund Council.

van der Grinten, T. 1996. "Scope for Policy: Essence, Operation and Reform of the Policy of Dutch Health Care." Pp. 135-154 in *Fundamental Questions about the Future of Health Care,* edited by L.J. Gunning-Schepers, G. J. Kronjee, R.A. Spasoff. The Hague: Sdu Uitgevers.

van der Grinten, T., P. Meurs, and K. Putters. 1998. "Healthcare Reform in the Context of Alliances: Social Entrepreurship in Dutch Health Care." Paper presented at the annual meeting of the European Healthcare Management Association, Dublin, Ireland, June 24-26.

van de Ven, W. 1997. "The Netherlands." Pp. 87-103 in *Health Care Reform: Learning from International Experience,* edited by C. Ham. Buckingham: Open University Press.

Versluis, J. 1993. *Management Participation by Medical Specialists in General Hospitals.* [in Dutch] Utrecht, Netherlands: Nationaal Ziekenhuisinstituut.

THE WORK OF GENETIC
CARE PROVIDERS:
MANAGING UNCERTAINTY AND AMBIGUITY

Kristen Karlberg

ABSTRACT

Prenatal genetic testing is fast becoming standard practice in the medicalized arena
of pregnancy in American health care provision. The interest of this paper, using
empirical research data from participant observation and semistructured interviews
of genetic counselors, geneticists, perinatologists, and obstetricians, is to explicate
the provision of genetic care by the care-givers themselves, paying close attention to
the ways they deal with the inherent uncertainties and ambiguities in medical genetics, especially prenatal genetic testing. Ambiguity and uncertainty are omnipresent
in prenatal genetic testing, most obviously through the absence of an individual to
examine in conjunction with test results. The test is for *fetal* abnormalities. Rarely
are test results able to be interpreted with a clear, straightforward definition of what
type of individual the fetus could eventually be. Through analysis of genetic intake
meetings, departmental meetings, and quarterly interdepartmental meetings, the
way providers order their work is elucidated; it reveals two work ideologies implemented to handle ambiguity and uncertainty: assessing the patient and tailoring the

Research in the Sociology of Health Care, Volume 17, pages 81-97.
Copyright © 2000 by JAI Press Inc.
All rights of reproduction in any form reserved.
ISBN: 0-7623-0644-0

information to the patient. These work ideologies are examined through a social worlds/arenas theory and a sociology of work lens informed by symbolic interactionism. Analyzing providers' interpretations of their clinical practices allows an explication of their (re)construction of genetic medical knowledges through the individual providers' social worlds.

THE INNOVATION OF GENETIC CARE

There is so much new in genetics and in ultrasound techniques that it seems like we are finding a lot of things that we don't necessarily know absolutely what it means prognostically. (Genetic Counselor)

I think technology has kind of messed things up. (Perinatologist)

You can never take medicine and standardize it like you can an assembly line. There is not a single fastest way to counsel a patient. (Medical Geneticist)

Through the findings of the Human Genome Project and concurrent research on specific genetic diseases, prenatal genetic care is today a burgeoning area of genetic practice in the United States. There are two common techniques for prenatal genetic testing—amniocentesis (amnio) and chorionic villus sampling (CVS)—and one for prenatal genetic screening—expanded alpha-feto protein testing (AFP). Amniocentesis is an invasive procedure that removes a sample of amniotic fluid from the uterus by drawing it through a needle inserted into the pregnant woman's abdomen, guided by ultrasound. The test can be performed at 15 to 20 weeks of pregnancy with a 1% chance of miscarriage (California Genetic Disease Branch 1995). CVS is done at nine to 11 weeks and is performed either transabdominally, by inserting a needle through the pregnant woman's abdomen into the uterus to extract a sample of the villi surrounding the fetus, or transcervically, which involves inserting a long plastic catheter through a woman's cervix into the uterus and withdrawing a sample of the villi. With CVS there is a 1% to 3% miscarriage rate (California Genetic... 1995). Both diagnostic tests are over 90% accurate in diagnosing chromosomal abnormalities (California Genetic... 1995). Testing for abnormalities in a gene or genes rather than in chromosomes (genes are the smaller units) must be requested at the time of the procedure, as the standard test culture only examines the material at the chromosomal level. AFP is a noninvasive blood test given to women at 15 to 20 weeks gestation, providing an indicator as to whether a woman is at higher risk than other women her age, weight, and gestational age for three anomalies: Down Syndrome, Trisomy 18, and neural tube defects (California Genetic... 1995). The test has a 25% false positive rate (Haddow et al. 1994). If a woman receives a positive test, indicating there may be something wrong with her fetus, her options are (1) amniocentesis to detect any chromosomal anomaly, (2) detailed ultrasound to detect neural tube defects, and/or (3) waiting for delivery. If she receives a negative test, it does not mean her child is "normal"[1], it merely means the likelihood of her child having one of the birth defects screened for with the AFP blood test is "normal" compared to other women her age, weight, and gestational age. Prenatal genetic testing is dramatically changing the experience of pregnancy for the woman, her partner, and her family as well as for providers (for example, see Kolker and Burke 1994; Rothenberg and Thompson 1994; and Rothman 1989).

These technological innovations of prenatal diagnosis are in the process of becoming routine in medical practice. The diagnostic testing technologies of amniocentesis and chorionic villus sampling coupled with screening techniques such as the expanded alpha-feto protein test are now the recommended standard of care for managing[2] high risk pregnancies (Pearson 1974), while in some states expanded AFP screening is mandated to be *offered* to all pregnant women. Even though diagnosis of genetic disease is impossible to standardize because of the ambiguity of genetic information obtained from amnio and CVS, many reputable scientists, including molecular biologist and Nobel laureate Walter Gilbert (1992, p. 94), believe genes can explain much of what medicine has been attempting to harness for centuries: "...a whole variety of human susceptibilities will be recognized as having genetic origins". The translation from genetic information provided by these technologies to genetic diagnosis using Gilbert's "genetic origins" supposition lies in the hands of a burgeoning professional segment (Bucher 1988), genetic care providers.

This paper examines the work of genetic care providers through their own eyes, using empirical data from participant observation and semistructured interviews. Despite the inherent ambiguity of genetic information and the heterogeneity of situated knowledges of different professional groups, I found that genetic care providers order their work though many work ideologies. Two discussed here are assessment work and tailoring of information work. These work ideologies enable genetic care providers to "processually" order (Strauss 1993) the ambiguities of genetic information and their professionally situated knowledges and to frame the work they do every day, managing the abstract science of genetics to provide medical diagnoses. The focus is specifically the way the uncertainties and ambiguities of genetic diagnoses, which are diagnoses with uncertain prognoses, are ordered in the practice of genetic care through work ideologies. This research is founded on symbolic interactionism, which informs social worlds/arenas theory and the sociology of work related to the management of uncertainties.

THE FOUNDATION OF THE ANALYSIS

Social Worlds/Arenas Theory

Social worlds/arenas theory concerns the organization of life. A *social world* is a structural unit within which negotiated social orders are constructed and reconstructed (Clarke 1990). An *arena* "includes all collective actors committed to acting within it" (Clarke 1991a, p. 128). In the arena, various issues are debated, negotiated and manipulated by representatives of the worlds or sub-worlds (Strauss 1978). All social worlds theory is specific to the collective action taking place within and with the arena in question. Prioritizing the relativity and "situatedness" of knowledges (Haraway 1991) and allowing for diversity of such knowledges are foundational to my approach using social worlds/arenas theory. Arenas and social worlds are fluid, contingent, and permeable.

Sociology of Work

The sociology of work, specifically scientific work, is not the study of professions or occupations, but an examination of the actual work conducted through concrete tasks and activities, treating the work itself as collective action. Strauss' (1993) "processual ordering" was elaborated from Strauss and colleagues' (1964) work on negotiated order, which proposes that structure and order emerge through interaction. Processual ordering considers the "collective working out of ordering, involving self-interactive actors and the various interactive processes…a theory of acting rather than a theory of action" (Strauss 1993, p. 258). Processual ordering does not obliterate the importance of negotiation in the ordering of structures but instead deepens the understanding of negotiated orders through its emphasis on the temporality, immutability, instability, and degree of flexibility of the interactants. Star's (1983, p. 210) outline of scientific work analytic premises is useful here: (1) understanding work is accomplished neither by examining its products (2) nor by understanding it without its products; (3) work is interactive and processual and (4) meaning is not inherent to scientific work but is (re)negotiated continuously by workers and consumers.

The (re)negotiations are especially useful when dealing with ambiguities and uncertainties. Star and Griesemer (1989) assert that "science requires cooperation—to create common understandings, to ensure reliability across time, space and local contingencies", and genetic care is not excluded. One mechanism employed in cooperation and negotiation is simplification (Star 1983), which can effectively erase "chains of inference" in scientific work. The most common type of uncertainty genetic care providers face is "technical uncertainty" (Star 1985) through the limitations of data available to them concerning the interpretation and clinical significance of test results. Davison, Macintyre, and Smith (1994, p. 345) have outlined aspects of predictive genetic testing that produce uncertainty, and some of these are applicable to prenatal genetic testing: "variations in modes of inheritance, multi-gene conditions, the variable expression of genetic material, interaction with environment,…and the issue of accuracy and 'quality control' in testing." The work ideologies of assessing the patient/family and tailoring the needed information to the patient/family are negotiation sites for the inherent ambiguities and uncertainties of genetic care by genetic care providers.

PREVIOUS STUDY OF GENETIC CARE PROVIDERS

Little attention has been paid to the perceptions of genetic care providers about their work. Many if not most studies focus on women who were offered prenatal genetic testing/screening, attempting to elucidate the women's experiences during the process and the effects of the process on them and their pregnancies, thus including providers in a tertiary manner. Rona and colleagues (1994) found that genetic counseling after the birth of an affected child did not aid in understanding the genetic component of the disease and even caused a deterioration in social circles. Nelkin (1996) explored the cultural context of genetic screening and testing founded on genetic determinism, citing genetic explanations, which are used as a "way out" for parents with troubled children. Nelkin problematizes the fact that genetic testing is seen as a way to predict and control life. Rapp (1998), in concert with her earlier work on miscommunication in genetic care interactions between providers and patients based on multicultural misinterpretation by providers (1997), reveals that refusal of prenatal diagnosis does not reflect a misunderstanding of scientific information but is often a multicultural experience–based decision incorporating religion, kinship, social support, or reproductive history issues and concerns. Rapp (1998, p. 68) also presents patients as "moral pioneers" and "ethical gatekeepers vis-a-vis this technology."

Studies specifically of providers are less common. Work on providers mainly examines outcomes measured by patient data, but it does provide a lens through which to view this study: the routinization of genetic care into prenatal care is increasing. Bernhardt and Bannerman (1982) examined referral patterns for amniocentesis among obstetricians and found that obstetrician board certification indicated high referral, and Catholic religious orientation was highly related to nonreferral for the procedure. Press and Browner (1993) argue that providers gave women information about the expanded alpha-feto protein test in such a way as to convince them the test was something they *should* have rather than providing them with a "neutral" presentation and allowing the women to make their own informed decisions. Richards and Green (1993) found that the routinization of prenatal diagnosis into antenatal care by genetic providers causes women who refuse amniocentesis to be bracketed as "abnormal". Though prenatal genetic testing is innovative in the provision of care for pregnant women, the delivery and access issues remain troublesome.

Research focused on the way genetic care providers organize their work is scarce. One study interviewed providers, addressing ethical issues surrounding genetic testing, but questioned primary care providers rather than those trained for genetic care provision (Geller and Holtzman 1995). Another group of studies (e.g., Wertz 1993; Wertz and Fletcher 1989; Wertz and Fletcher 1998; Wertz, Fletcher, and Mulvihill 1990) conducted over a period of nearly 10 years examined the perceptions of medical geneticists from 37 nations on a number of ethical issues related to prenatal testing. One of these findings is of great interest: "Worldwide, 94–98% of the respondents would disclose ambiguous or conflicting prenatal test results to patients" (Wertz 1993, p. 7). For the United States, 75% of medical geneticists reported they would disclose colleague disagreement about ambiguous or "artifactual" test results, while 89% would disclose new or controversial interpretations of test results (Wertz 1993, p. 27).

Yet another study (Kerr, Cunningham-Burley, and Amos 1998, p. 194) interviewed clinicians and scientists working in the "new genetics", which the authors define as involving "research into the genetic components to a range of disease, illness, and behavior and its application in the clinic in the form of testing, screening and treatment." They found that genetics researchers believed there were two main differences between eugenics and the new genetics: (1) new genetics is considered to be based on individual informed choice, not coercion, and (2) the new genetics is based on claims concerning a combination of nature (heredity) and nurture (environment) as opposed to nature only.

THE STUDY

Genetic Care Providers and Their Work at a Managed Care Facility

This paper is based on empirical research using participant observation and qualitative interviews. The research took place at a managed care facility, a not-for-profit health maintenance organization (HMO) in Northern California. Genetic services within the multifacility HMO are coordinated by a group of medical geneticists who both provide genetic care and plan for future services by working with the HMO administration. My research was focused at one site, interacting with individuals primarily from the Genetics and Perinatology Departments. I attended seven Genetics Department meetings lasting two hours each. I also sat in on three interdepartmental meetings, each one and a half hours long, with representatives from radiology, obstetrics, perinatology, and genetics present. Handwritten notes were taken during the meetings and later transcribed. I also interviewed providers: six female genetic counselors, two male medical geneticists, two male perinatologists, and a male obstetrician who performs prenatal diagnosis procedures. The interviews ranged in length from twenty minutes to an hour and 20 minutes and were conducted in the interviewee's offices or in a departmental conference room. The interviews were tape recorded and transcribed. The data were collected between November 1997 and March 1998. Grounded theory was used in data collection and analysis (Glaser and Strauss 1967), which has its foundation in symbolic interactionism (SI).

THE ARENA

HMO-Centered Prenatal Genetic Testing

The focus of this analysis is the work of the providers, including their perceptions of interactions with their patients. The genetic care providers' work consists of negotiations among the actors involved in prenatal diagnosis in the HMO: professionals in radiology, obstetrics/perinatology, and genetics; the medical records chart room staff; the patients and their partners, families, and friends; and the wider HMO bureaucracy. Here the research emphasizes the negotiations between the primary genetic care providers: medical geneticists, genetic counselors, and perinatologists. The pregnant women whom the negotiations revolve around are invisible actors in these interactions, but they are very present in the data. Second-hand accounts of what the women said and felt were common from the providers in both meetings and interviews.

The analysis has three main parts. First the situated knowledges of the three different professions involved in genetic care are explicated: genetics, genetic counseling, and perinatology. Then, the shared work practices of assessment and tailoring of information are examined through the data. Finally, the situated knowledges and ambiguities of genetic information are revealed through the data of the shared work practices.

Situated Knowledges of the Genetic Care Providers

The three medical departments involved in this HMO-centered prenatal diagnosis arena are radiology, obstetrics/perinatology and genetics. Radiology and obstetrics are not represented in this sample because their work is tertiary to the provider–patient interactions of prenatal testing[3]. As the work of these genetic care providers is under the umbrella of the HMO, there is a prescribed protocol or pathway, which all women must follow when prenatal diagnosis is deemed appropriate and chosen. If a pregnant woman is considered of normal risk and will be at least 35 at the time of delivery, if she is referred by an obstetrician after a prenatal visit, or if she is considered high risk, she is offered prenatal genetic testing. If she chooses to have testing, there are three major steps she will follow at this HMO: (1) she meets with a genetic counselor; (2) the actual testing is performed by a perinatologist; and (3) the test results are delivered, explained, and options are discussed. Step three is handled by the genetic counselor if the results are "normal," but if they are "abnormal," usually a genetic counselor and a medical geneticist meet and discuss the options with the woman and her partner.

Genetic Counselors

Genetic counselors are masters' level trained in a program emphasizing molecular genetics and medical counseling. Genetic counselors meet with pregnant women and explain the tests available, the risks involved, and the options after testing. If women choose to have testing, counselors schedule the test and are present the day of the test. Genetic counselors also deliver results and meet with women and medical geneticists, if necessary. Genetic counselors view patient advocacy as one of their primary roles. One genetic counselor told me, "I feel like we are definitely a resource. We are one of the last holdouts in medicine where we can still meet with people for an hour or longer. If they're really upset we can stay with them all day." The profession of genetic counseling is nearly all female. Genetic counseling serves as an "auxiliary occupation, [a] deliberately fashioned lower status group" to genetics (Bucher 1988, p. 143). The counselors do work that the medical geneticists used to do—work that became routine to the specialty and so was relegated to genetic counseling: medical record chart review, initial meetings with patients, family history taking, and, in this HMO's pathway for prenatal testing, actually explaining and offering the testing.

Perinatologists

Perinatologists are subspecialists who attended medical school, were granted a residency in obstetrics and gynecology, and then earned a subspecialty in perinatology. They are specifically trained to manage high risk pregnancies and are schooled in treating both mother and fetus until delivery. A perinatologist described the difference between his role and those of medical geneticists: "We take care of clinically what goes on on the labor floor while the [medical] geneticists deal with the cerebral issues of what the prognosis is, but they (medical geneticists) don't really participate in clinically managing the problem." The perinatologist's role in genetic care provision is performing the prenatal diagnosis procedures of amniocentesis and chorionic villus sampling. With an "abnormal" result, they manage pregnancies if women choose to continue. While medical geneticists and genetic counselors only follow women who have had the diagnosis of a genetic disease or some equally ambiguous genetic finding, the perinatologists have contact with the patients throughout their pregnancies.

Medical Geneticists

Medical geneticists are medical doctors who did a residency in a specialty such as pediatrics, internal medicine, or obstetrics and then a specialty in genetics. They diagnose and manage genetic diseases. A medical geneticist who has been in practice at this HMO for thirty years described his job as, "not only challenging, but a tremendous variety of interesting things...so I can stay interested in what I do." Another medical geneticist called his practice, "very nonroutine." One medical geneticist explained the most difficult facet of practicing genetics:

What is most difficult is seeing a child or a fetus where you know there is something wrong but you cannot make a diagnosis. So you have to deal with the uncertainty of what that is. I can prove to the parents, or I can point out to the parents where it might even be better not to know the diagnosis. But it's still better probably in the long run to know the diagnosis, they just might not want to know what the eventual out come will be.

The main roles of the medical geneticists in this social arena of prenatal diagnosis are the interpretation of results from the tests to make a diagnosis and management following the diagnosis.

IDEOLOGY: SHARED WORK PRACTICES

In the data around the actual collective practice of genetic care giving, the providers (regardless of specialty) viewed their work of producing and managing genetic diagnoses as composed of the overarching tasks of (1) assessing the patient/family and (2) tailoring the needed information to the patient/family. Each of these tasks is a situation where negotiation and renegotiation of genetic care providers' jobs in the arena of HMO-centered prenatal diagnosis occur regularly.

Assessment Work–Assessing the Patient/Family

Assessing the patient enables the providers to practice more "competent" care. There were five major issues the genetic care providers suggested were useful to their assessments. Patient assessment includes determining the following:

1. knowledge level about genetic topics
2. apparent comprehension of the information about the testing available for prenatal diagnosis
3. whether a prochoice or anti-abortion (prolife) stance is favored
4. the comfort level when dealing with uncertainty and ambiguity
5. a broad sketch of the patient's personality type

In most cases, the assessment is performed by genetic counselors in their initial visits with patients by asking them questions and observing overall attitudes. The providers had various interpretations of the same patient. Much of the interpretation was based on social position including such variables as education level, religiosity, and other relative measures. A genetic counselor noted one patient she met with brought her pastor's wife with her to the meeting and was clasping a Bible on her lap throughout the session. To this genetic counselor, her observations meant the woman was not going to consider abortion if something "abnormal" was found.

Assessment work involves evaluating patients on subjective perceptions and assumptions of genetic care providers. An obstetrician, acting as a perinatologist, recounted a case that reflects his interpretation of patient comprehension and knowledge level:

> ...A teenager with this really malformed baby with her teenage support system, and it creates a sort of difficult situation. This baby, if it survives, will be really retarded most likely. They were sort of blithely streaming along...and there didn't seem to be any recognition.

This informant has disdain for the woman in question, based on her age and possibly her socioeconomic status. Another instance was recalled by a genetic counselor:

> I must say I've certainly met couples who I know have been counseled extensively and went through a hole...this woman had had a positive AFP result[4]. She'd come in for follow-up, and there had been a whole new dating of her pregnancy with the screening test reinterpreted to normal. She went on with her pregnancy and had her baby, who was normal. She's still worried about what will appear in her baby that will explain this positive screening test. I'm sure the counselor talking to her assumed that there was a level of understanding. Yet I'm sure there was a degree of unawareness about what her level of comprehension was. And I think that often happens. We're dealing with such complicated stuff every day. You begin to think some of it seems easier than you realize it is to the person who's facing it for the first time.

The genetic counselors showed an enormous amount of compassion for the women and the complex experience of prenatal genetic testing. Patients' personalities were focal points for genetic counselors in their assessment work. Defining a personality was a hybrid of patients' social positions and their capability of maneuvering in the maze of genetic information. Patient personality and dealing with uncertainty are often linked issues, as illustrated in this case by a genetic counselor.

> I had a woman with an AFP of 10.5 and normal average is 1. And we couldn't say that this baby wouldn't make it to term or have some sort of syndrome that would make it mentally retarded. We had to really give her the range of outcomes in that sort of situation, and she had a really hard time with it. And I think part of it was the situation; it was just so ambiguous. But it was also her personality. She was very focused on finding answers. And she got many different opinions, but nobody could tell her the right thing to do. She wondered how we could not have the answer. And she ended up not continuing....She felt guilty about terminating, but she knew she did not want to deal with a baby that had multiple problems. We didn't have any guarantees that that would be the case either, so it was much more difficult.

Counselors expressed frustration themselves with the grey areas in the diagnosis of genetic tests.

This epitomizes the inherent ambiguity of genetic testing in general but especially prenatal genetic testing. For prenatal genetic testing, no individual can be examined in conjunction with the test results, as the test is for *fetal* "abnormalities". A genetic counselor reflects on this problem:

But it's telling them [the diagnosis] and asking them to make a judgement and decision on that condition, the particular condition before they've ever had a chance possibly to deal with what we're talking about. They don't have that child there...

The diagnosis rests on the medical geneticists' knowledge and experience, as well a her/his confidence in presenting the diagnosis as "certain". Rarely are there "clear-cut " diagnoses. Even when the diagnosis is "straightforward" such as with a fetus diagnosed with Down Syndrome having three copies of chromosome 21 instead of the "normal" two copies, the level of metal retardation and physiological manifestations of the disease are not quantifiable from prenatal genetic testing results.

The patient assessment, examining conditions and contingencies that can affect the course of the interaction, is a complex process involving the genetic care providers ordering their work around the ambiguities and uncertainties of genetic diagnoses. The laundry list of issues thought to impact the way a woman receives the information about genetic testing or the results from such testing qualifies the structure of the provider–patient interaction. It is a mini-processual ordering (Strauss 1993). The ambiguity of test results makes this work task extremely important because of the uncertainty, which must be conveyed to most patients with a genetic diagnosis. Once the patient assessment has been conducted, the information relevant to that patient's case is tailored based on the feelings and pieces of personal information gleaned from previous interactions.

Tailoring Information Work—Tailoring the information for the Patient/Family

The providers anticipate being able to tailor the information about a prenatal diagnosis procedure being considered or a genetic disease or birth defect being diagnosed.

Going [in to meet with] a woman who's carrying a baby with a problem, just the knowledge of whether I know that she was considering abortion or not changes how I approach the situation. Having that information ahead of time makes it easier to develop an approach, a strategy. Now I've been fooled before, and I've had to change in midstream when I've realized that I've made the wrong decision or got the wrong information, and I've misread the patient. And I have to sort of back-pedal and start all over again. (medical geneticist)

Tailoring the information involves presenting the options available, such as the various tests, abortion, adoption, and the multiple ways of managing the pregnancy, in a specific way to fit the patient's individual needs. A genetic counselor said:

There's so many differences [in the patients] because there's differences in diagnoses, there's differences in family's attitudes, there's difference in couple's attitudes, in couple's relationships to their pregnancy, relationships to feelings about abortion, about the types of decisions they would make.

Such tailoring of the information may also include a specialized description of the disease. A disease description could highlight the positive or negative outcomes associated with the disease while minimizing the opposite. In ironic contrast to the possibility of specializing a disease description based on the patient, full disclosure of diagnosis information was favored by the providers. The providers overwhelmingly stated that their policy was to not withhold any information:

I believe patients should be told as much as we know. I don't withhold information (perinatologist)

Our policy…is that they're gonna know everything that we know about their child or child-to-be whether it's good or bad (medical geneticist).

This is in line with the providers surveyed by Wertz, Fletcher, and Mulvihill (1990). While the genetic care providers believed they were being forthcoming with diagnosis information, the way they represented this tailoring information work reveals that, in subtle ways, the information a patient receives is not "all" of the information available.

One medical geneticist explains his tailoring approach[5]:

…I try to make a judgement call, I'm probably right more than I'm wrong, but it's still a judgement call, about what sort of information they need… So it's possible that somebody, two different families with children having the same condition may get different spins based upon what their perceived needs are…if somebody's in denial to the point of that creating perpetual problems, … I may swing some verbally, to try and get them to come to the reality that you can't deny this, that your child has this…The other parent may be so focused on the negative side that I'll spin very positive….So the way I say it will sound very different. But I'm trying to bring them to what I think is the correct place—something that's intellectually correct.

This "intellectually correct" position is very subjective, but a common goal of medical geneticists and perinatologists/obstetricians. The "correct" position is founded on the diagnosis of the particular provider, so it may be different based on who is the diagnosing medical geneticist. The ambiguity of results and the impossibility of a guarantee of a "perfect baby" even with the "normal" genetic testing results negate the concept of a "firm" diagnosis of test results. Genetic counselors problematize their approach to the situation:

The possibility that I always grapple with is that we give people too much information that they may not be prepared to deal with, and that sounds a little patronizing and I don't mean it that way. One of the points of counseling people, especially before prenatal procedures, is getting some sense of what this might mean to them. To get an idea so that that can just help you frame things. When you have anything with grey information, because it is grey so you do have to try and play both sides, but more so because people are looking for it to be black and white. Not one wants to be in a grey area. So they're going to take out of that

one version. That's human nature to make sense. No one wants to deal with mushy, grey,
ambiguous information. (genetic counselor)

The process of presenting the information in a patient-specific manner is the
epitome of negotiation in this arena. The genetic care providers' jobs change
depending upon the patient (s)he is dealing with, as the job skills vary with each
individual. All of the social worlds in the HMO-centered prenatal genetic testing
arena affect the outcome of the negotiation; the patient assessment is altered based
on the patients and providers and on each of their beliefs, feelings, and ideologies,
all of which are shaped by their work, home and social environments (for
example, see Ritchey, Yoels, Clair, and Allman 1995).

Medical geneticists and genetic counselors generally follow the "nondirective"
tenet, which asserts that the providers should not direct patients to make any
particular decision in response to a genetic diagnosis. Genetic care providers are
supposed to present the information in a manner that is nondiscriminatory and to
allow patients to make the decision based on a "neutral" presentation of the data.
A medical geneticist described his perspective of his practice overall:

> So I'm seeing a pretty select group of people who for the most part have made their decision
> that if anything's wrong with this kid they're not going to continue [the pregnancy]. On the
> other hand there is a small group who will change their mind based on the degree of severity
> of the condition and even a smaller group who are having the amnio just to find out if there's
> anything wrong so they can prepare for that. And I just have to know what group they fit in, or
> what counseling is going to give them the most support and benefits. So although we still say
> we're nondirective we do—are directive—in the fact that we might tailor the counseling for
> what we know the family is going to do no matter what we tell them.

Providing genetic care is different than general medical practice, as is evidenced by this research. In general medical practice, the physician makes a diagnosis based on the findings in a medical examination or test and suggests a course
of action for treatment of the problem. The suggested course of action is usually
followed by patients. Genetic care by medical geneticists and genetic counselors
does not allow for suggestions in most cases because the providers are to present
information and allow the patient to make an "informed" decision. The way of
practicing genetic care "nondirectively" is a challenge for the providers.

CONFLICTING CONCLUSIONS

Where Ambiguities and Situated Knowledges Collide with Ideologies of Shared Work Practices

The overarching ideology common to the HMO-centered prenatal genetic testing arena is the belief in the ambiguity of genetic information. It is a situated
knowledge, which impacts the application of the shared work practices. This is a

"gray area" of diagnosis, when there is a genetic "abnormality", but not one typified by a medical definition or known medical significance. Ambiguous information is part of the rush and challenge of practicing genetic medicine; the ever-changing diagnosis, the new finding. But ambiguity is frustrating for most patients and can be emotional turmoil for some. The providers recognize this instability and its impact on patients and incorporate it into their work ideologies.

The mode of practice regarding directiveness is another ideology made more complex by the data. This study demonstrates the directive/nondirective dichotomy is not a clear division. It is nearly impossible to practice genetic care in a nondirective manner because the genetic information is loaded with uncertainty. The very nature of the work tasks revealed through the data is subjective. Based on an interpretation of patients, the providers make decisions regarding what information is best suited to the diagnosis those patients receive. As pure nondirectiveness is not possible, the dichotomy no longer exists. Problematizing nondirectiveness is not new (e.g., Clarke 1991b). The idea of "non-directiveness" must be further interrogated as prenatal genetic testing is becoming more common[6].

Interestingly, my data led me to the conclusion that the boundary imposed by Kerr and colleagues (1998) between eugenics and the new genetics was blurred. In this HMO facility, the practice of genetics does not allow for environment. This is evident by the power of the genetic diagnosis, regardless of how uncertain, to alter perceptions of patients and fetuses. Nature, the actual genes and chromosomes of the fetus, was the basis of the diagnosis made. Environment as nurture was not allowed for in the diagnosis. Also, although informed choice is a goal of genetic care provision, it is many times not achieved. This is seen through the genetic counselors' awareness of the lack of patient comprehension despite high education level and other variables expected to aid understanding. So while Kerr and colleagues (1998) suggested genetics was moving toward a more enlightened approach informed by choice and nondeterminism or, at the very least, a multifactorial model, my research shows that while this is still a noble goal, it does not appear easily implemented. Most of the providers thought choice and nondeterminism was the proper approach, but they were not certain it was attainable.

The genetic care work tasks represented in the arena of an HMO-centered prenatal diagnosis discussed here are assessment work and tailoring information work. These tasks performed by genetic counselors, medical geneticists, and perinatologists/obstetricians are influenced by the interactions and negotiations taking place in the social and work lives of both the providers and patients. In practice, genetic health care providers process the information the patients give them and provide information to the patients in a way these providers believe is "tailored to the patient." The social worlds–arenas theory informed by the sociology of work enables the explication of the various situated knowledges of the different specialties interacting in the work of the providers. As Clarke (1991a, p. 19) has stated and this research supports, "social worlds characteristically generate *ideologies* about how work should be done..." and, in this instance, the genetic care providers manage uncertainties through patient assessment and tailoring information based on that assessment.

ACKNOWLEDGMENT

I wish to thank the members of the Genetics Department at the HMO where this research was conducted for their openness and willingness to share their work with me. I also wish to gratefully acknowledge the following for their thoughtful and encouraging comments on earlier drafts of this paper (in alphabetical order): Diane Beeson, Adele Clarke, Melanie Egorin, Jennifer Fishman, Jennifer Fosket, Jennie Jacobs Kronenfeld, Laura Mamo, Virginia Olesen, Teresa Schertzer, and Janet Shim.

NOTES

1.   I use quotes around the words "normal" and "abnormal" to problematize their usage generally in regard to genetics and specifically relating to the fetus and its genetic material.

2.   "Managing" pregnancy is a term used by physicians to refer to treating a pregnant woman throughout her pregnancy. It is usually used in cases of "high risk" pregnancies, where decisions have to be made about the best course of action for the pregnant woman and her fetus in order to obtain the most desirable outcome for both woman and fetus. I see this term as a manifestation of power over pregnant women by physicians because the female patient is seen as something to be steered through the course of pregnancy.

3.   Women have only one or two encounters with Radiology throughout the prenatal genetic testing process and are not in direct interaction in most cases. The radiologist, who is a medical doctor, conducts routine ultrasound for women at approximately 19 weeks of gestation. If something "abnormal" is found, a genetic counselor is called in and the prenatal diagnosis protocol ensues. Obstetricians refer high risk women or women over the age of 35 at delivery to the genetic counselors for prenatal diagnosis counseling. They do not play a role other than referral in prenatal genetic testing, with the exception of the obstetrician in this sample who performs the prenatal genetic testing procedures. He was the only obstetrician interviewed for this study. No radiologists were interviewed for this study. Time was also a consideration in their exclusion. These roles are prescribed by the protocol of this HMO social world, dictated by practice guidelines and other organizational constraints.

4.   Refer to "The Innovation of Genetic Care" section for a definition of AFP.

5.   This medical geneticist worked mainly with children who were diagnosed with a genetic disease after birth rather than fetuses diagnosed *in utero*.

6.   Meaney, Riggle, and Cunningham (1993) found that the number of amniocenteses performed jumped from 8,700 in 1980 to 74,000 in 1989.

REFERENCES

Bernhardt, B., and R. Bannerman. 1982. "The Influence of Obstetricians on the Utilization of Amniocentesis." *Prenatal Diagnosis* 2: 115-121.
Bucher, R. 1988. "On the Natural History of Health Care Occupations." *Work and Occupations* 15(2):
131-147.

California Genetic Disease Branch. 1995. "The California Expanded AFP Screening Program." and "Prenatal Testing Choices for Women 35 Years and Older." pamphlets produced by the California Department of Health Services, California Genetic Disease Branch, Berkeley, CA.

Clarke, A.E. 1990. "A Social Worlds Research Adventure: The Case of Reproductive Science." Pp. 15-42 in *Theories of Science in Society*, edited by S. Cozzens and R. Gieryn. Bloomington, IN: Indiana University Press.

_____. 1991a. "Social Worlds/Arenas Theory as Organizational Theory." pp.119-158 in *Social Organization and Social Process: Essays in Honor of Anselm Strauss*, edited by D. Maines. New York: Aldine De Gruyter.

Clarke, A. 1991b. "Is Non-Directive Genetic Counseling Possible?" *Lancet* 338(October 19): 998-1001.

Davison, C., S. Macintyre, and G.D. Smith. 1994. "The Potential Social Impact of Predictive Genetic Testing for Susceptibility to Common Chronic Diseases: A Review and Proposed Research Agenda." *Sociology of Health and Illness* 16(3): 340-371.

Geller, G., and N.A. Holtzman. 1995. "A Qualitative Assessment of Primary Care Physicians' Perceptions About the Ethical and Social Implications of Offering Genetic Testing." *Qualitative Health Research* 5(1): 97-116.

Gilbert, W. 1992. "A Vision of the Grail." Pp. 83-97 in *The Code of Codes: Scientific and Social Issues in the Human Genome Project,*edited by D. Kevles and L. Hood. Cambridge, MA: Harvard University Press.

Glaser, B., and A. Strauss. 1967. *The Discovery of Grounded Theory: Strategies for Qualitative Research*. New York: Aldine De Gruyter.

Haddow, J.E., G.E. Palomaki, G.J. Knight, G.C. Cunningham, L.S. Lustig, and P.A. Boyd. 1994. "Reducing the Need for Amniocentesis in Women 35 Years of Age or Older with Serum Markers for Screening." *New England Journal of Medicine* 330(16): 1114-1118.

Haraway, D.J. 1991. *Simians, Cyborgs and Women: The Reinvention of Nature*. New York: Routledge.

Kerr, A., S. Cunningham-Burley, and A. Amos. 1998. "Eugenics and the New Genetics in Britain: Examining Contemporary Professionals' Accounts." *Science, Technology and Human Values* 23(2): 175-198.

Kolker, A., and B.M. Burke. 1994. *Prenatal Testing: A Sociological Perspective*. Westport, Connecticut: Bergin and Garvey.

Meaney, F.J., S. Riggle, G.C. Cunningham. 1993. "Providers and Consumers of Prenatal Genetic Testing Services: What do the National Data Tell Us? " *Fetal Diagnosis and Therapy* 8(Suppl 1): 18-27.

Nelkin, D. 1996. "The Social Dynamics of Genetic Testing: The Case of Fragile-X." *Medical Anthropology Quarterly* 10: 537-550.

Pearson, J. 1974. "The Management of High-Risk Pregnancy." *Journal of the American Medical Association* 229: 1439-1440.

Press, N., and C. Browner 1993. "Collective Fictions: Similarities in Reasons for Accepting Maternal Alpha Fetoprotein Screening among Women of Diverse Ethnic and Social Class Background." *Fetal Diagnosis and Therapy* 8 (Suppl 1): 97-106.

Rapp, R. 1997. "Communicating About Chromosomes: Patients, Providers and Cultural Assumptions." *Journal of the American Medical Women's Association* 52(1): 28-29.

_____. 1998. "Refusing Prenatal Diagnosis: The Meaning of Bioscience in a Multicultural World." *Science, Technology and Human Values* 23(1): 45-70.

Richards, M.P.M., and J. Green 1993. "Attitudes Toward Prenatal Screening for Fetal Abnormality and Detection of Carriers of Genetic Disease: A Discussion Paper." *Journal of Reproductive and Infant Psychology* 11: 49-56.

Ritchey, F.J., W.C. Yoels, J.M. Clair, and R.M. Allman. 1995. "Competing Medical and Social Ideologies and Communication Accuracy in Medical Encounters." *Research in the Sociology of Health Care* 12: 189-211.

Rona, R., R. Beech, S. Mandalia, D. Donnai, H. Kingston, R. Harris, O. Wilson, C. Axtell, A. Swan, and F. Kavanagh. 1994. "The Influence of Genetic Counseling in the Era of DNA Testing on Knowledge, Reproductive Intentions and Psychological Well-Being." *Clinical Genetics* 46: 198-204.

Rothenberg, K., and E. Thomson, Editors. 1994. *Women and Prenatal Testing: Facing the Challenges of Genetic Technology.* Columbus, OH: Ohio State University Press.

Rothman, B.K. 1989. *Recreating Motherhood: Ideology and Technology in a Patriarchal Society.* New York: W.W. Norton.

Star, S.L. 1983. "Simplification in Scientific Work: An Example from Neuroscience Research." *Social Studies of Science* 13: 205-228.

_____. 1985. "Scientific Work and Uncertainty." *Social Studies of Science* 15: 391-427.

Star, S.L., and J. Griesemer. 1989. "Institutional Ecology, 'Translations' and Boundary Objects: Amateurs and Professionals in Berkeley's Museum of Vertebrate Zoology, 1907-39." *Social Studies of Science* 19: 387-420.

Strauss, A.L. 1978. "A Social Worlds Perspective." pp.119-128 in *Studies of Symbolic Interaction,* edited by N. Denzin Greenwich, Connecticut: JAI Press.

_____. 1993. *Continual Permutations of Action.* New York: Aldine de Gruyter.

Strauss, A.L., L. Schatzman, R. Bucher, D. Ehrlich, and M. Sabshin. 1964. "Negotiated Order and the Co-ordination of Work." pp.175-202 in their *Psychiatric Ideologies and Institutions.* New York: The Free Press.

Wertz, D.C. (1993) "International Perspectives on Ethics and Human Genetics." 27 *Suffolk University Law Review* 1411 (from Westlaw database).

Wertz, D.C., and J.C. Fletcher. 1989. "An International Survey of Attitudes of Medical Geneticists Toward Mass Screening and Access to Results." *Public Health Reports* 104(1): 35-44.

_____. 1998. "Ethical and Social Issues in Prenatal Sex Selection: A Survey of Geneticists in 37 Nations." *Social Science and Medicine* 46(2): 255-273.

Wertz, D.C., and J.C. Fletcher, and J.J. Mulvihill. 1990. "Medical Geneticists Confront Ethical Dilemmas: Cross-Cultural Comparisons among 18 Nations." *American Journal of Human Genetics* 46: 1200-1213.

BIOMEDICAL MARKERS, ADHERENCE MYTHS, AND ORGANIZATIONAL STRUCTURE: A TWO-STAGE MODEL OF HIV HEALTHCARE PROVIDER DECISION MAKING

William D. Marelich, Oscar Grusky, Jeff Erger, Traci Mann, and Kathleen Johnston Roberts

ABSTRACT

This study developed a two-stage model of decision making used by healthcare providers when considering combination antiretroviral therapy for HIV-positive patients. Provider interviews and patient medical records from 10 HIV-positive patients were utilized. Four providers (two nurse practitioners and two physicians' assistants) situated in two managed care HIV clinics were studied. Newly diagnosed HIV patients were observed while interacting with their health care provider. Following the interactions, healthcare providers were given a semistructured private interview about the factors that influenced their treatment decisions. It was hypothesized that providers, when they made their treatment decisions, would initially

Research in the Sociology of Health Care, Volume 17, pages 99-117.
Copyright © 2000 by JAI Press Inc.
All rights of reproduction in any form reserved.
ISBN: 0-7623-0644-0

evaluate biomedical markers followed by an evaluation of patient characteristics associated with treatment adherence. It was found that providers developed basic decision-making principles and perceived some decisions as relatively easy and others as much harder. The easiest decisions conformed to biomedical guidelines while the harder ones had substantial behavioral and nonmedical components. These were hard decisions because they were perceived as making treatment regimen adherence much more problematic. It was also found that the providers' high level of decision-making responsibility combined with their relatively low authority contributed to their decision-making uncertainty.

INTRODUCTION

The intent of this study was to examine the factors influencing the decision-making process used by healthcare providers when prescribing combination antiretroviral therapy to Human Immunodeficiency Virus/Acquired Immune Deficiency Syndrome (HIV/AIDS) patients. Although a great deal is known about individual and organizational decision making in such diverse settings as juries, aircraft cockpits, and university departments (e.g., Janis and Mann 1977; March 1994; Weick 1979; 1990), relatively little is known about nonmedical factors that influence the treatment decisions of healthcare providers (c.f., Haynes, Taylor, and Sackett 1979). Decisions made about disease treatment affect the daily lives of those with HIV/AIDS. In particular, decisions regarding drug treatments for those with HIV/AIDS may influence not only disease progression and the choice of future treatments but also quality of life and psychosocial issues. For example, a strict drug regimen of *saquinavir* for the HIV-positive individual, which requires taking six pills three times daily with a large meal, can severely limit activities the individual can perform. Further, for the patient who is trying to conceal his/her HIV status, disclosure of this treatment regimen may lead to problematic psychosocial issues such as loss of social support from peers, friends, or family members (Herek 1990).

Understanding this decision-making process in treating those with HIV/AIDS is especially important given the changing healthcare environment and how care is administered. For example, healthcare organizations are increasing their service loads (i.e., number of patients seen and treated) but keeping the same number of healthcare providers to conserve costs. This increase in provider/patient ratio limits the amount of time providers can spend with their patients, and these time-limitations can affect decisions (Pfeffer 1985). Further, healthcare organizations are making greater use of nonphysician professionals (i.e., nurse practitioners or physicians' assistants) to diagnose and treat patients and, for case-management, in an effort to conserve costs (Day 1999). As with physicians' decision making in prescribing medication, little or no empirical research exists on how these ancillary healthcare professionals make their decisions. Thus, as healthcare organizations move toward increased patient loads and utilization of ancillary healthcare professionals for treatment, the question of how treatment decisions are made becomes paramount.

The role of the healthcare provider is that of a decision maker (Israël 1982). Physicians and other primary health providers (e.g., nurse practitioners, physician assistants) may be viewed as managers of patient health who oversee almost all levels of care for the patient. Hence, these providers have much in common with managers in many organizations where decisions influence outcomes. For example, an auto assembly plant manager uses quality control information, speaks with line employees, tours the plant daily, and consults with other managers (e.g., those in operations, engineering, materials) to help inform his/her daily decisions about production and staffing needs (Terkel 1975). In a similar vein, healthcare providers use their knowledge of medicine, disease progression, treatment options, relevant medical guidelines, and professional consultation (e.g., with other healthcare providers) to provide appropriate care for the patient.

However, uncertainty about treatment outcomes can interfere with the providers' treatment decisions (Israël 1982). This may be especially true for HIV healthcare providers in prescribing combination antiretroviral therapy with protease inhibitors. Although combination antiretroviral therapy employing protease inhibitors can greatly increase the lifespan and quality of life of the patient (DHHS 1997), the complex and potentially toxic regimens present difficulties in patient adherence. Moreover, if patients are unable to comply with the strict drug regimens, they are at risk for developing a drug-resistant strain of HIV (Carpenter et al. 1997).

In making their treatment decisions for those with HIV, healthcare providers evaluate biomedical markers such as T-cell and viral load laboratory test results. Providers also assess adherence cues based on associated beliefs and myths. This evaluation and assessment is performed within an organizational structure that may ultimately influence providers' decisions. Biomedical markers are often quantitative in nature. On the other hand, adherence is murky because (a) adherence behavior is inherently unclear (Haynes, Taylor, and Sackett 1979), (b) healthcare providers tend to be poor judges of patients' adherence (Gordis 1979), (c) adherence expectations are often constructed from inadequate provider/patient communications and hence are unrealistic, and (d) judgements of the probability of nonadherence may be based on entities that co-vary but are not causally related. Furthermore, providers fill a role in their organization and therefore conform to organizational pressures and structures that can constrain their decisions.

Adherence (also called compliance) has been defined as "the extent to which patients follow the instructions—proscriptions and prescriptions—of their physicians or other providers" (Haynes, Taylor, and Sackett 1979, p. 63), although what actually constitutes successful and unsuccessful adherence is not clear. Adherence rates for medications taken over long periods tend to converge around 50% (Sackett and Snow 1979), indicating that patients in these situations take their prescribed medication about half the time, and rates of 75% to 80% are considered successful adherence (Horwitz and Horwitz 1993). However, Gordis (1979) has noted that simple adherence rates (such as those above) can have different meanings. For example, an adherence rate of 50% could indicate a patient is taking his/her daily prescribed medication every other day, while another patient may follow a cycle of skipping three days and then taking his/her medication for three days. In both cases, patients' adherence is 50%.

Studies in HIV/AIDS care have shown that missed doses of antiretroviral therapies are strongly associated with increases in patients' viral load levels (Chesney 1997; Vanhove, Shapiro, Winters, Merigan, and Blaschke 1996). However, it is not yet known what the optimal amount of adherence is for those with HIV or even how best to define adherence in this domain (e.g., missed doses, doses taken at the wrong time, etc.). Indeed, Roberts (1998) recently showed that even "expert" HIV/AIDS physicians do not know "how much adherence is enough" for these complicated antiretroviral regimens.

Contributing to the murky concept of adherence are *adherence myths*. Myths in social organizations are "constructed to provide broad explanations of life and models for behavior" (March 1994, pp. 208-209) and are assembled with information and stories the decision maker has acquired (Mahler 1988). Knowledge of medical research, stories from colleagues, and information acquired in medical training all can contribute to the overall shape and character of myths. For example, research suggesting that adherence rates are associated with patient factors such as unemployment, homelessness, current or past abuse of substances, fatalistic attitudes, or low income (Baekeland and Lundwall 1975; Broers, Morabia, and Hirschel 1994) may be woven into the fabric of adherence myths. Myths are a collective story and show consistency from person to person in an organization (Mahler 1988) much as myths of creation of the world show great similarities across cultures.

When interacting with patients, healthcare providers cast the individual patients into archetypal roles that become part of the myths surrounding adherence. These patient roles are constructed through the communications between the healthcare provider and the patient, where the standards and consequences of nonconformity with drug regimens are generated from both sides. In the same way certain archetypes are present in ancient myths (e.g., hero, trickster; see Campbell 1949), the general medical archetypes are the good patient/bad patient. Within each of these categories are specific subtypes such as drug addict or mentally ill. It is important to note that not all patients are cast into these roles, even if the characteristics of the role are present. One provider might always see recreational drug users as unlikely to adhere, whereas another provider might see some drug users as good adherence risks. What matters most is how the archetype is made to fit with the overarching adherence myths. Another provider might question why a particular patient is likely or unlikely to adhere (given information consistent with a specific archetype), but they will not question the nature of the adherence myth itself.

We see the influence of adherence myths on patient treatment as unfolding in the following way. Myths of adherence exist for the provider and are created and conditioned by the environment of the organization and the provider's career experience. Each individual patient is cast into a specific archetypal role early in the treatment process, and this role-casting is done based on a few cues that emerge in the interaction (for an analysis of such cues, see Erger, Grusky, Mann, and Marelich in press). At the same time, adherence myths are communicated to the patient. The healthcare provider then uses the archetypal label to justify the treatment decision in terms of adherence myths to both the patient and other members of the healthcare organization.

Adherence myths, as interpreted by providers, therefore operate as causal cues (Einhorn and Hogarth 1986), which are indicative of whether patients will succeed or fail in taking their prescribed treatment regimens, and are supported and reinforced as part of the organizational culture (March 1994). However, adherence myths can be misleading (Montagu 1997). Just because patients fit a particular profile does not mean that they will not adhere to a strict treatment regimen. Although providers are responsible for setting the expected standards for adherence, they often have little understanding of what are realistic expectations because of incorrect information concerning adherence and lack of knowledge about the patients' situation (Haynes, Taylor, and Sackett 1979). Patients, on the other hand, are expected to conform to these idealized standards, although they are sometimes not informed fully about what the standards are and what the consequences of nonconformity may be (Mann, Grusky, Marelich, Erger, and Bing in press). Further, indicators that went into constructing the myths (e.g., medical research, stories from colleagues, etc.) may simply co-vary with no causal link (Einhorn and Hogarth 1986).

Beyond adherence myths, the structure of the organization in which the decision maker is embedded can influence heathcare providers' decision making (Pfeffer 1985). Social structures such as organizations strive for stability and predictability and therefore establish paradigms of operation (i.e., defined roles, rules, hierarchies, negotiated order, and shared meanings) to ensure stability (Brown 1978; Weick 1979). Individuals within the organization become generalized, fitting an organizational role and following set rules (Wiley 1988). Healthcare providers may be seen as part of an organizational environment, embedded within the fabric of the organizational structure and bounded by roles and hierarchies. In turn, this bounded environment can influence treatment decisions. For example, in a related study, Erger and colleagues (in press) found that organizational pressures for providers to enroll new patients in clinical trials influenced treatment decisions. This type of pressure is analogous to ecological pressures and resource dependencies (e.g., costs, efficiencies, personnel, legal and fiscal barriers, political constraints) that can influence organizational decisions and outcomes (Pfeffer 1985; Vaughan 1996).

For the current investigation, we propose that provider decision making may be best understood as a two-stage model where providers, within a bounded organizational environment, initially try to maximize patient care by prescribing antiretroviral therapy for patients if they meet biomedical guidelines (thereby focusing on biomedical markers) and then evaluate patient characteristics for adherence. Prescribing or deferring antiretroviral treatment is the final decision point. If biomedical guidelines are met and the patient characteristics do not co-vary with nonadherence issues, then combination antiretroviral therapy with protease inhibitors is provided. However, if patient characteristics are associated with nonadherence, then alternative therapy strategies may be adopted. The latter option does not suggest that patients receive subpar care but instead receive treatment that the provider believes to be more appropriate in light of anticipated nonadherence.

In sum, this study examined the application of this two-stage model of decision making used by HIV healthcare providers when considering combination antiretroviral therapy for HIV-positive patients. Provider interviews and patient medical record abstracts from 10 HIV positive adults were used. Newly diagnosed HIV patients were observed while interacting with a healthcare provider. After the interactions, healthcare providers were interviewed and asked to "walk through" how they reached their decision to prescribe or defer combination antiretroviral therapy to the observed patients using a semistructured interview protocol. A qualitative analysis using multiple readings of transcripts (Strauss and Corbin 1990) was utilized to investigate emergent themes from the interviews.

METHOD

Participants

Participants were 10 HIV patients (nine male) ranging in age from 28 to 50 (M = 34.5). All patients were on public assistance, and four patients were African-American, five were Caucasian, and one was Latino. To be eligible for the study, patients had to (a) never have received combination antiretroviral therapy for HIV, (b) be 18 years of age or older, (c) be new patients at the clinics under study, (d) be asymptomatic, and (e) speak enough English so that their interactions with the provider would be in English.

The patients were seen by one of four female healthcare providers. Two of the providers were nurse practitioners and two were physicians' assistants. The providers worked at one of two managed care clinics whose clients were limited to HIV patients. We intended to have physicians participate in the study, but both clinics primarily used nurse practitioners and physicians' assistants for first-time HIV patient assessment and treatment. Whether this is normative for similar clinics is unknown, although it is increasingly likely that managed care healthcare institutions use nonphysician providers as "front-line" personnel for initial diagnosis and treatment (Day 1999). Clinic A served an urban, ethnically diverse population (60% of clients receive Ryan White benefits; 35% Caucasian, 16% African-American, 48% Latino, 1% Asian/Pacific Islander). Clinic B served a more affluent population (50% of clients receive Ryan White benefits; 50% Caucasian, 23% African-American, 20% Latino, and 7% Asian/Pacific Islander).

There were a total of five full-time primary care providers at Clinic A that serve approximately 540 patients per month, with 40 new patients per month. Two of the providers included in this study see approximately 50% of clinic A's patients between them. Both physicians' assistants work at Clinic A. For Clinic B, there were a total of three full-time primary care providers, and the two providers included in this study see approximately 60% of Clinic B's patients between them. Clinic B providers serve approximately 800 patients per month and receive 30 new patients per month.

The providers for the current study are responsible for the medical care of 200 to 400 patients at any given time, with new patients added monthly. One provider noted "Well, I carry a load of about 400 [patients] altogether, you know, that comes and goes. So the number I've treated [with antiretrovirals] is very high...." The clinics allocate approximately 15 to 30 minutes for the provider to meet with the patient. For the current study, meeting times ranged from 8 to 40 minutes with a mean of 26.4 minutes.

Procedures

Eligible patients were invited to participate in the study by clinic receptionists. If patients were interested, two researchers described the study to them and obtained their consent. Providers gave consent to participate before the researchers approached patients, and were given a copy of the "Guidelines for the Use of Antiretroviral Agents in HIV-Infected Adults and Adolescents" (DHHS 1997) approximately two months prior to the observations and interviews.[1]

During the observation stage of the study, two observers accompanied each patient to an examination room where the interaction took place (for details on this protocol, see Mann et al. in press). The interaction was a conversation and hence did not involve a medical examination or medical tests. After the observation stage, the observers and patient left the room. Interviews with the provider were conducted by one of the researchers later that same day (typically right after the medical visit). Interviews lasted approximately 20 minutes and used a semistructured format. All interviews were audiotaped and transcribed for analysis.

Provider Interview Protocol

The provider interview was open-ended and designed to elicit information from each provider on how they came to a decision regarding treatment for each patient and on what factors were important to them in reaching that decision. Providers were first asked about their general impressions of the patient and about their relationship with the patient. After citing the treatment decision for each patient, providers were asked to list all the factors that they considered in coming to that conclusion. Thirteen factors were listed on the interview protocol (patient chart, interaction with patient, ability of patient to pay, adherence, clinic guidelines, other guidelines, other staff opinions, drug use history, psychiatric history, job situation, domestic situation, education level, and connection with clinical trials) and providers were asked to elaborate on each factor they mentioned as important. They were also asked if they ever considered those factors from the list that they had not mentioned and to elaborate on those factors.

Analysis

Following the suggestions of Strauss and Corbin (1990), multiple readings of the transcripts were performed to identify major ideas or themes in the participants' descriptions of their situations. As new themes emerged from each transcript, the previously analyzed transcripts were reevaluated for the new themes. Important and frequently mentioned ideas were grouped into coding categories—for example, regimen adherence, deferment, and healthcare provider decision-making sequence. Matrices were constructed to help store and manage the data during the analysis process (Miles and Huberman 1994).

RESULTS

Overall, five of the 10 patients were provided combination antiretroviral therapy. A summary of patient laboratory results is shown in Table 1. Of the five treated cases, three met both DHHS (1997) criterion guidelines of less than 500 CD4+ T cells/mm and more than 10,000 (bDNA) copies of HIV RNA/ml of plasma, while two of the cases met only the less than 500 CD4+ criterion. Initially, this suggests that providers are following the DHHS guidelines to provide antiretroviral therapies to patients based on whether they met one criterion or both criteria for treatment.

Two patients, however, met the DHHS standards for therapy (Cases 1 and 5), but were deferred antiretroviral therapy. Case 1 met the criterion of less than 500 CD4+, while Case 5 met both T-cell and viral-load criteria. The decision to defer therapy seemed surprising given Case 5's high viral load and low T-cell count. In comparison, Case 7 was offered treatment, yet had T-cell and viral load figures that were less extreme on both criteria. Taken as a whole, these findings show that patients are sometimes deferred treatment despite clearly meeting the DHHS medical standards, indicating that providers may consider information other than the DHHS medical marker standards when deciding whether or not to recommend therapy to patients.

Table 1.   T-Cell and Viral Load for 10 Observed Patients by
Provider Type and Clinic

Case	Provider	T-Cell	Viral Load	DHHS Guideline Interpretation	Provider Decision
Clinic A					
1	PA 1	355	870	Treat	Defer
2	PA 1	156	500	Treat	Treat
3	PA 1	968	1502	Defer	Defer
4	PA 2	508	4688	Defer	Defer
5	PA 2	301	>30000	Treat	Defer
Clinic B					
6	NP 1	361	8496	Treat	Treat
7	NP 2	480	11840	Treat	Treat
8	NP 1	860	513	Defer	Defer
9	NP 1	203	21000	Treat	Treat
10	NP 1	131	>30000	Treat	Treat

Note: NP is "Nurse Practitioner", PA is "Physicians' Assistant.

Table 1 also shows that four of the five cases observed in Clinic A were deferred while four of five cases observed in Clinic B were treated by prescribing antiretroviral agents. The two clinics differed with regard to the type of professional training and experience of the providers. Clinic A was staffed by physician's assistants while Clinic B was staffed by nurse practitioners. Both groups were supervised by physicians and both were part of the same larger managed care organization. Although there was a trend in the direction of differences between Clinic A and Clinic B, the differences shown in Table 1 between the treatment decisions of these two groups were not statistically significant ($p = 0.198$, Fisher's Exact Test, 2-tailed).

Provider Decision Making

Several assumptions guided our analysis of provider decision making. First, we assumed that rationality is limited in the decision-making process, as Simon (1960; 1997) noted in his discussions of "bounded rationality." Rationality is limited because the organization or social system is unable to provide maximum or even adequate information to enable the decision maker to make fully informed decisions. At the same time, however, decisions are not random. We anticipated that providers would use medical information to guide their evaluations of patients. What makes decision making difficult is uncertainty. We expected that providers would seek to alleviate or reduce uncertainty by relying on their judgments and experiences and on their organizations. These influences act as constraints on the decision process.

Providers adapt to the decision-making requirements on their job by developing basic principles that they follow, what we call *the 1, 2, 3's of provider decision making*. Consequently, they are likely to perceive some of their decisions as relatively *easy decisions*, while others are much harder. The easiest decisions are likely to be those that conformed almost fully to biomedical guidelines. The harder ones are likely to be cases that have substantial behavioral and nonmedical components, such as drug abuse and serious psychiatric diagnoses, which are likely to make *treatment regimen adherence* much more problematic.

Finally, we anticipated that the providers' role in the organization, that is, their high level of decision-making responsibility combined with relatively low authority, would contribute to decision-making uncertainty. Providers were either physicians' assistants or nurse practitioners and were under the supervision of medical doctors. They were "in the middle" in the sense that they presented themselves to patients as experts, yet their authority was considerably less than that of a physician. The consequence of this was the tendency to *attribute responsibility* for the treatment decisions to the patient. We turn next to the four themes addressing decision making and accountability that emerged from the analysis: *the 1, 2, 3's of provider decision making, "easy" decisions, regimen adherence and deferment,* and *attributing responsibility.*

The 1, 2, 3's of Provider Decision Making.

To enable understanding of the decision-making process for their treatment decisions, providers were asked "Why is this the appropriate treatment for this patient? In other words, could you walk me through the information you used to come to this decision, what you considered first, second, and so forth?" Seven of the 10 case reports contained three primary decision-making steps taken to decide upon a particular treatment. For example, the provider for Case 5 reported:

> I initially [*first*] looked at his T-cell count and his viral load, and [*second*] I told him that yes, he definitely wanted treatment. However, I felt he's not a candidate to start treatment until his bipolar disease [*third*] was stabilized.

Of the 10 case reports, seven noted "lab results" as the first bit of information considered. This underscores the importance of the providers using biomedical markers in making treatment decisions. However, the above case also mentions patient motivation or willingness to start antiretroviral medication, a theme noted in eight of the 10 cases. For example, the healthcare provider for Case 2 reported:

> [*First*]...[I] knew he was willing to [take on the regimen], and interested in taking the medication, because not everybody is. And...number *two*, by looking at his lab results. And number *three*, I thought that he could handle going through the side effects, and not everybody can.

For this case, patient willingness was considered first, followed by lab results and side-effects. In addition, this was the only case that mentioned side-effects as part of the treatment decision. It should be noted that this provider saw a total of three cases, and lab results and patient willingness were apparent across all three of the cases. Another example of the use of biomedical markers and patient willingness in making the treatment decision was provided by Case 9:

> His T-cell count and his viral load, *first* off. And also…I would definitely recommend it simply because I believe it can improve his immune system. And [*second*] he says he's ready to start, he's extremely ready to start today. He's definitely motivated, wants to save his life, he…told us that. His numbers, his immune system, the HIV-markers, his readiness. He seemed to be agreeable that he definitely was motivated and that he could do it, and he would do it.

A recurrent theme in the decision-making sequence was patient adherence, which was mentioned in five of the 10 cases. As anticipated, adherence plays a critical role in the final treatment decision. For example, the provider for Case 4 reported:

> I think number *one*, I really felt like he wasn't ready, even if he really needed the medication—say if he had really low T cells and a high viral load…he's just not ready to start medicine.… The *second* concern was the drug use. And I know that when you're using crystal, which he's doing, you're staying up for two or three days at a time, and you're not gonna take your medicine, so…I don't want to set him up for resistance or failure…and because *thirdly*, he's not in the danger zone in terms of immune function or viral load. So I felt very comfortable in waiting…some time.

For this case, lab results were mentioned first, followed by adherence concerns and then noting that the patient was not in the "danger zone." This case was on the borderline for meeting the DHHS treatment criteria, with a T-cell count just over 500 and a viral load under 5000. According to the Guidelines, this case could have been treated if the provider and patient had decided that was an acceptable route. However, the provider's concern about the patient's drug use seems to have "tipped" the decision for deferment.

"Easy" Decisions

Overall, nine of the 10 cases were given a definitive treatment decision, while one case (Case 7) needed additional laboratory results at the time of the observations (a few days later, new lab results were obtained and the case was treated with antiretrovirals). Providers were asked "What was your treatment decision for the case we observed?" Providers gave ample information that substantiated their treatment decision, ranging from the patient meeting established medical guidelines for treatment to the patient's immediate need for a "strong" antiretroviral therapy. For example, the provider who deferred antiretroviral treatment for Case 3 noted medical criteria, "He had very high T cells, 900 range, and…although he had a viral load, it was…very low…so I felt like we had more time to play with [for] him." In another example, treatment deferment was also based on laboratory findings. The provider for Case 8 reported:

> It was a pretty easy decision because of the…national guidelines…his T-cell count and viral load. I mean, I really didn't devaluate him if I thought maybe he would be able to adhere to a regimen…. So, I guess it was nothing about him actually, except his T-cell count and viral load.

For cases in which antiretroviral treatment was recommended, providers appeared to have relied on the patient's "need" for therapy, thereby suggesting that the lab results met the medical criteria for treatment, and now was the time to find an appropriate HIV therapy. For example, the provider for Case 6 noted "I just chose what we feel is the best first HIV [combination] therapy, and leaving options for him, because it seemed like there were no counterindications to that." In another example, the provider for Case 2 noted the following:

> I thought that [combination therapy] was appropriate for that patient because I felt like he needed…a strong combination, a very powerful combination. So, giving him two protease, and two nucleosides…. I had to look at other parameters, like his kidney, liver, and in this case, this gentleman was anemic, I wouldn't have given him AZT. And I wouldn't have wanted to give him a triple-combination…that's not as powerful, because of the advanced HIV disease.

Overall, these results suggest that when patients met established biomedical markers, treatment or deferment of antiretroviral therapy was frequently an easy decision.

Regimen Adherence and Deferment

When severe threats to treatment adherence arose, even though cases met the appropriate treatment criteria, some cases were deferred antiretroviral treatment. The two cases that met DHHS guidelines but were not treated showed clear evidence for adherence concerns. Case 1 was deferred because the patient was using recreational drugs, while Case 5 was deferred for a psychiatric condition. In each case, these two reasons were linked to antiretroviral drug adherence. For example, the provider for Case 5 reported:

> The treatment was to wait [defer], even though I think he's a candidate for antiretroviral therapy. His T-cell count was around 300, and his viral load was about 30,000 to 40,000. So that indicates to me that he definitely needs some antiretroviral therapy…but he does have a serious bipolar disease, and the likelihood of him being able to comply with medicines, given that he's not on any medication and he's not stable, was probably zero to none. Slim to none…. So, what I decided was to hold off on that [antiretrovrial treatment] for a while, and to treat the little complaints that he had, which was some of the skin problems, and to offer to get him hooked up with a psychiatrist.

This treatment decision contrasts with a comment made later in the same interview that is indicative of the provider's general treatment philosophy: "It's recommended that you start treatment on anybody regardless of their T-cell count if you have a high viral load." Although this provider's statement indicates that all cases should be treated if the viral load meets DHHS standards, the provider's former statement clearly shows that factors beyond medical criteria play an important role in the evaluation of the patient's medication adherence.

Case 1 was also deferred antiretroviral therapy due to adherence concerns. Although the provider noted that the patient was willing to try the treatment regimen, she decided to defer the patient due to the patient's drug use history and current drug use. In the words of the provider for the case, "My fear [is] that as she…starts using drugs more that she won't be able to be compliant…she'll be high, she'll forget the doses, she'll be out on the street."

Reasons given for deferment of the remaining three cases were varied. Case 8 was deferred because he did not meet the biomedical criteria. Although Cases 3 and 4 did not meet DHHS guidelines, their healthcare providers noted reasons for deferment other than medical criteria. Case 3 was deferred because the patient did not want to go on antiretroviral therapy, "I asked him how he felt about taking medications, and…[it] was pretty clear that he did not want to [take the medications] right now, and that he had a lot of concerns about taking medications." Case 4 was deferred for drug use, "The treatment decision was to not start antiretrovirals at this time, based on the fact that he had recently been using drugs, he'd probably not comply with that medication." These examples suggest that although providers could have simply used medical guidelines to justify deferment, instead they provided reasons associated with treatment adherence or patient motivation.

Overall, these results regarding deferment of antiretroviral treatment suggest that providers focus on adherence concerns when making decisions about prescribing antiretroviral therapy. In particular, these concerns center around the patient's lifestyle or personal characteristics, including recreational drug use, patient motivation, and psychiatric problems.

Attributing Responsibility

Providers made broad attributions that the final treatment decision was not their responsibility, with seven of the 10 case reports noting that the treatment decision was ultimately "up-to-the-patient." For example, the provider for Case 6 concisely stated "Even though we have all these guidelines and recommendations, it's ultimately up to the patient."

In another example, the provider for case 10 reported "I felt like he was a good candidate for it [treatment], certainly, because he was basically telling me, you know, I want to start it." The provider for Case 1 provided a longer explanation:

And sometimes I tell people—You know, this is my opinion, and this is what I think, but you have to do what you want—and I do do that...[I give them]...guidance. But it's not my job to make their decisions.

For three of the 10 cases, providers did not attribute the final treatment decision to the patient. For example, the provider for Case 8 noted, "I guess it was nothing about him actually, except his T-cell count and viral load," thereby attributing her decision to the biomedical markers to defer the patient antiretroviral treatment. In another example, the provider for Case 5 attributed her decision to an adherence myth, "I felt he's not a candidate to start treatment until his bipolar disease was stabilized."

Having 70% of the case reports attributing the decision to the patient is surprising considering the import of the combination antiretroviral decision with its complicated regimen and sometimes harsh side-effects. It is generally accepted that patients should have a say in their medical treatment. At the same time, Mann and colleagues (in press) found in a related study that it is rare that HIV healthcare providers fully informed their patient of the risks (or even benefits) associated with antiretroviral medication.

One explanation for attributing the treatment decision to the patient stems from the organizational status of the provider (Erger et al. in press). The status of physicians' assistants and nurse practitioners is low compared with physicians, who make the same types of treatment decision. This combination of low status with high responsibility generates uncertainty in making the treatment decision. Thus, high uncertainty in the organization, combined with low authority and high responsibility, may lead to the tendency of attributing the decision to the patient.

DISCUSSION

The findings in this study suggest that healthcare providers considered two basic factors in prescribing combination antiretroviral therapy—laboratory results and patient regimen adherence—thus confirming the hypothesized two-stage model of provider decision making. Providers generally considered lab results first, followed by an assessment of the likelihood of adherence based on patient characteristics. Easier decisions were those where lab results clearly conformed to biomedical guidelines and patients did not have characteristics stereotypically associated with nonadherence, while harder decisions were those where nonmedical components required closer evaluation. Indeed, two of the 10 cases were deferred treatment although they qualified under the DHHS Guidelines.

Patients who presented themselves as motivated, confident, knowledgeable, and "ready-to-go" were viewed by providers as good candidates for antiretroviral therapy. If they met medical guidelines for treatment, and did not have a history of recreational drug use or psychiatric problems, this type of individual was provided antiretroviral therapy. However, if there was an indication of nonadherence (which may be called "red-flagging" the patient), then providers satisficed (Simon 1997) or settled for a "good enough" treatment decision in an attempt to isolate and stabilize problematic characteristics. In each case, providers noted that another evaluation would be performed in the future for the patient. This reevaluation process served as a decision feedback loop (Janis and Mann 1977) giving the provider the option to reevaluate the treatment decision at a later time.

Table 2 summarizes the hypothesized two-stage model and additional components that can affect provider decision making. In addition to nonphysicians, this table also contains information on how the model might generalize to physicians. Although we did not collect information from physicians, the two-stage model presumably applies as well to physicians with some notable differences. The roles imposed by the *organizational structure* are represented by the nonphysician and physician categories. Both roles are high in responsibility, yet nonphysicians have lower status in the organization. The *two-stage process model* suggests that providers evaluate biomedical markers, followed by adherence issues. Biomedical markers may be considered a "clean" evaluation because of their apparent quantitative nature based on biomedical information used as markers for high-risk cases. Adherence issues, which are murky at best, are given either low or high relevance to the decision based on the prior biomedical information. Since these issues are based on myths and the good/bad patient archetype, evaluation of adherence relevance can result in unrealistic expectations of patient adherence. *Outcome uncertainty* refers to the equivocal or unequivocal nature of the treatment decision to provide antiretroviral medication. These decisions may be high or low in uncertainty based on who makes the decision (e.g., physician or non-physician), the biomedical markers, and the adherence evaluation.

These findings suggest that treatment adherence is a strong motive in providers' decisions to treat patients with antiretroviral therapies or to defer. However, as noted earlier, the issue of adherence is rather equivocal. The techniques providers use to make sense of adherence are analogous to the techniques individuals use to discern an ambiguous picture. For example, when an individual is presented with an ambiguous picture (e.g., a fuzzy picture of a four legged animal), he/she

Table 2. Factors Affecting Decision Making and Uncertainty

| Org. Structure (Provider Role) | Process | | Outcome Uncertainty |
	Stage 1: Biomedical Markers	Stage 2: Adherence Import	
Nonphysician (e.g., NP, PA)	High T-Cell Count, Low Viral Load	Low	High (hard decisions)
Physician (M.D.)	Low T-Cell Count, High Viral Load	High	Low (easy decisions)

attempts to clarify the picture by attending to a greater number of cues (e.g., the animal is on a beach, a person is next to the animal, the animal is smaller than the person) to make the final interpretation (e.g., the fuzzy picture is a dog). In our sample, healthcare providers used myths about adherence and patient archetypes as cues to reduce ambiguity. However, since these myths are constructed through social interactions between the provider and the patient, and are influenced by the existing organizational structure, they often may contain incomplete and inaccurate information due to organizational malcommunication.

Furthermore, time pressures and provider fatigue may have had an effect on the role and evaluation of adherence myths in the decision process. As noted before, our providers had case loads as high as 400 patients. Time and fatigue often limit or interrupt the decision-making sequence, thus leading to a "stopping rule" that forces a final decision without a thorough evaluation of the problem (Pfeffer 1985). The effect of this stopping rule may have been reflected in the handling of two cases that qualified for treatment according to the DHHS guidelines but were deferred.

For seven of the 10 cases, the final treatment decision was attributed to the patient. This, as noted earlier, may be a function of the low status of the heathcare providers in their organization coupled by their high responsibility for decisions. In addition, given the retrospective nature of the interview setting, these decisions and patient-oriented attributions may be couched as a form of sense making (Weick 1979). The framework of sense making suggests that actions or outcomes (such as treatment decisions) occur first, followed by an attempt to make sense or attribute the action or outcome to some perceived cause.

Sense making and its focus on "thinking" or consideration after action is most closely linked to the theory of cognitive dissonance (Festinger 1957), which specifies that actions may be interpreted after the fact and evaluated for consonance or dissonance. In addition, research on attribution processes suggests that cognitive sequences such as sense making typically function to balance conflicting attitudes (Bem 1974). Hence, in applying sense making to this study, the physicians' assistants and nurse practitioners made their decisions first, which could have been riddled by high uncertainty if patient adherence was in question. Because of this high uncertainty in the decision, these "providers in the middle" reinterpreted their treatment decisions as a function of what the patient wanted, and therefore attributed the final treatment decision to the patient in order to reduce decision uncertainty.

There are several caveats associated with this study that should be mentioned:

1. The small sample size, both in regard to the number of patients observed and the number of healthcare providers interviewed, may limit the generalizability of these results to other providers and patients.

2. Physicians were not observed or interviewed; therefore, conclusions regarding differences between physicians and nonphysicians have not formally been tested.
3. Only two clinics within a large healthcare organization were sampled, which again may be problematic for generalization of results to other organizations or even clinics within the organization observed.

However, some strengths of the study include the opportunity to observe heathcare providers of different roles (e.g., nurse practitioners, physicians' assistants) in different ethnically diverse clinic settings that served a high number of HIV positive patients.

In summary, one of the main functions of the HIV healthcare provider is that of a decision-maker, who is asked to make easy and hard decisions in regard to treating patients with antiretroviral therapy. Indeed, our providers followed a similar decision-making sequence to those used by organizational managers, where individuals first appraised the problem, surveyed alternatives, deliberated about commitment to a final decision, then finally affirmed a decision choice (Janis and Mann 1977). We have suggested providers evaluate both biomedical information and information surrounding adherence to make their final treatment decisions. However, the decisions providers make are affected by their status in the organization, and by their reliance on myths surrounding the murky concept of treatment adherence. Uncertainty in the final treatment decision arises from these factors, and can lead providers to attribute treatment decisions to the patient.

Although this causal spiral appears to function adequately, our results suggest that some patients who qualify medically for antiretroviral medications are being deferred for issues related to adherence, a term not well conceptualized or understood. These patients are falling through the cracks of treatment, yet the decisions are argued as justified because of myths, stories, and suggestive research. One way to seal these cracks is to provide healthcare professionals with guidance in interpreting adherence issues and to help them improve their decision making.

As noted in the introduction, whether addressing HIV/AIDS or other disease treatment, little empirical work has been done on how healthcare providers make their treatment decisions. The two-stage decision-making model is a first attempt at understanding the many components associated with making treatment decisions. In this way, the model was designed to be portable for other diseases that are concerned with the effects of resistance to a drug regimen. For example, the model could be used effectively by healthcare providers who treat tuberculosis patients, especially given its similarity to HIV in the complexity of treatment considerations and in problems with resistant disease strains.

Decision-making models are constructed to aid our understanding of how and why particular decisions are made. As an end result, healthcare professionals and social scientists design interventions to make the decision process more efficient, accurate, and beneficial to patients. However, as with any explanative process model, the resulting decision-making improvements will have to "fit" the current system of healthcare (i.e., increased use of nonphysician providers and increased patient loads) and hence must be seen as an efficient addition to a system focused on quality of care and on cost containment. Indeed, the real challenge is to both improve, and implement, treatment decision-making processes.

ACKNOWLEDGMENT

Supported in part by NIMH grant MH19127 to Oscar Grusky.

NOTE

1.   Since the completion of this study, new guidelines have been released with slight modifications to recommendations (Department of Health and Human Services and the Henry J. Kaiser Foundation Panel on Clinical Practices for Treatment of HIV Infection. 1999).

REFERENCES

Baekeland, F., and L. Lundwall. 1975. "Dropping Out of Treatment: A Critical View." *Psychological Bulletin* 82(5): 738-783.

Bem, D. J. 1974. "Cognitive Alteration of Feeling States." Pp. 211-233 in *Thought and Feeling: Cognitive Alteration of Feeling States*, edited by H. London and R.E. Nisbett. Chicago, IL: Aldine Publishing.

Broers, B.A., Morabia, and B. Hirschel. 1994. "A Cohort Study of Drug Users' Compliance with Zidovudine Treatment." *Archives of Internal Medicine* 154:1121-1127.

Brown, R.H. 1978. "Bureaucracy as Praxis: Toward a Political Phenomenology of Formal Organizations." *Administrative Science Quarterly* 23: 365-382.

Chesney, M. 1997. "New Antiretroviral Therapies: Adherence Challenges and Strategies." Presented at the *Adherence in HIV Therapy: Implications for Nursing Practice Conference*, San Francisco, June 3.

Campbell, J. 1949. *The Hero With a Thousand Faces*. New Haven, CT: Yale University.

Carpenter, C., M. Fischl, S. Hammer, M.S Hirsh, D.M Jacobsen, D.A. Katzenstein, J.S. Montaner, D.D. Richman, M.S. Saag, R.T. Schooley, M.A. Thompson, S. Vella, P.G. Yeni, and P.A. Volberding. 1997. "Antiretroviral Therapy for HIV Infection in 1997." *Journal of the American Medical Association* 277(24):1962-1969.

Day, J.G. 1999. "Children and Health Care in the United States: Past, Present, and Future." Pp. 1-89 in *Children's Health Care: Issues for the Year 2000 and Beyond*, edited by T.P. Gullotta, R.L. Hampton, G.R. Adams, F.A. Ryan, and R.P. Weissberg. Thousand Oaks, CA: Sage.

Department of Health and Human Services (DHHS) and the Henry J. Kaiser Foundation Panel on Clinical Practices for Treatment of HIV Infection. 1997. *Guidelines for The Use of Antiretroviral Agents in HIV-Infected Adults and Adolescents*. November 5. Rockville, MD: HIV/AIDS Treatment Information Service Reproduction Center.

_____. 1999. *Guidelines for The Use of Antiretroviral Agents in HIV-Infected Adults and Adolescents*. May 5. Rockville, MD: HIV/AIDS Treatment Information Service Reproduction Center

Einhorn, H. J., and R.M. Hogarth. 1986. "Judging Probably Cause." *Psychological Bulletin* 99(1): 3-19.

Erger, J., Grusky, O., Mann, T., and W. Marelich. (in press). "HIV Health Care Provider/Patient Inter-action: Observations on the Process of Providing Antiretroviral Treatment." *AIDS Patient Care and STDs.*

Festinger, L.L. 1957. *A Theory of Cognitive Dissonance.* Stanford, CA: Stanford University Press.

Gordis, L. 1979. "Conceptual and Methodologic Problems in Measuring Patient Compliance." Pp. 23-45 in *Compliance in Health Care,* edited by R.B. Haynes, D.W. Taylor, and D.L. Sackett. Baltimore, MD: Johns Hopkins.

Haynes, R.B., Taylor, D.W., and D.L. Sackett (Eds.) 1979. *Compliance in Health Care.* Baltimore, MD: Johns Hopkins.

Herek, G.M. 1990. "Illness, Stigma, and AIDS." Pp. 107-150 in *Psychological Aspects of Serious Illness: Chronic Conditions, Fatal Diseases, and Clinical Care,* edited by P.T. Costa, Jr., and G.R. Vandenbos. Washington, D.C.: American Psychological Association.

Horwitz, R.I., and S.M. Horwitz. 1993. "Adherence to Treatment and Health Outcomes." *Archives of Internal Medicine* 153: 1863-1868.

Israël, L. 1982. *Decision-Making, The Modern Doctor's Dilemma: Reflections on the Art of Medicine,* translated by M. Feeney. New York: Random House.

Janis, I.L., and L. Mann. 1977. *Decision Making: A Psychological Analysis of Conflict, Choice, and Commitment.* New York: Free Press.

Mahler, J. 1988. "The Quest for Organizational Meaning: Identifying and Interpreting the Symbolism in Organizational Stories." *Administration and Society* 20: 344-368.

Mann, T., Grusky, O., Marelich, W.D., Erger, J, and E. Bing. (in press). The Implementation of the DHHS Guidelines for the Use of Antiretroviral Agents in HIV-Infected Adults: A Pilot Study. *AIDS Care.*

March, J.G. 1994. *A Primer on Decision Making: How Decisions Happen.* New York: Free Press.

Miles, M.B., and A.M. Huberman. 1994. *Qualitative Data Analysis.* Thousand Oaks, CA: Sage.

Montagu, A. 1997. *Man's Most Dangerous Myth: The Fallacy of Race, 6th ed.* Walnut Creek, CA: AltaMira.

Pfeffer, J. 1985. "Organizations and Organization Theory." Pp. 379-440 in *Handbook of Social Psychology: Volume I - Theory and Method* (3rd ed.), edited by G. Lindzey and E. Aronson. New York: Random House.

Roberts, K.J. 1998. "Adherence to Antiretroviral Treatment Regimens: Physician and Patient Perspectives." Unpublished doctoral dissertation, University of California, San Francisco.

Sackett, D.L., and J.C. Snow. 1979. "The Magnitude of Compliance and Noncompliance." Pp. 11-22 in *Compliance in Health Care,* edited by R.B. Haynes, D.W. Taylor, & D.L. Sackett. Baltimore: Johns Hopkins.

Simon, H.A. 1960. *The New Science of Management Decision.* New York: Harper & Row.

Simon, H.A. 1997. *Administrative Behavior: A Study of Decision-Making Processes in Administrative Organizations, 4th ed.* NY:Free Press.

Strauss, A., and J. Corbin. 1990. *Basics of Qualitative Research: Grounded Theory Procedures and Techniques.* Newbury Park, CA: Sage.

Terkel, S. 1975. *Working.* New York: Avon.

Vanhove, G., Shapiro, J., Winters, M., Merigan, T., and T. Blaschke. 1996. "Patient Compliance and Drug Failure in Protease Inhibitor Monotherapy." *Journal of the American Medical Association.* 276: 1955-1956.

Vaughan, D. 1996. "*The Challenger Launch Decision: Risky Technology, Culture, and Deviance at NASA.*" Chicago: University of Chicago Press.

Weick, K.E. 1979. *The Social Psychology of Organizing* (2nd ed.). New York: McGraw-Hill.

_____. 1990. "The Vulnerable System: An Analysis of the Tenerife Air Disaster." *Journal of Management.* 16(3): 571-593.

Wiley, N. 1988. "The Micro–Macro Problem in Social Theory." *Sociological Theory* 6: 254-261.

PART III

GENDER AND HEALTH

WOMEN'S HEALTH AND GENDER BIAS IN MEDICAL EDUCATION

Mary K. Zimmerman

ABSTRACT

Gender bias in medical knowledge and practice is an issue of longstanding significance for women's health scholars and activists alike. This paper assesses the current status of medical education, where gender bias has the potential to influence the culture and process of medical care, and focuses on three areas of concern: the presence and participation of women as medical students and faculty, the problem of gender bias in the content of medical curricula and training programs, and the friendliness for both men and women of the climate and environment of medical education. Significant change has occurred over the past several decades in the admission of women into medicine; yet, women remain under represented in positions of leadership and decision authority. In the 1990s, the content of the medical curriculum began to be evaluated in terms of gender, and a number of the resulting changes were implemented. Additionally, recent attention has been placed on improving the gender friendliness of medical school policies and resources. While these developments signal a decrease in gender bias and greater equality in medical education, the ability of the medical profession to continue to address these issues is being challenged by the increasingly powerful private health care economy.

Research in the Sociology of Health Care, Volume 17, pages 121-138.
Copyright © 2000 by JAI Press Inc.
All rights of reproduction in any form reserved.
ISBN: 0-7623-0644-0

INTRODUCTION

An important contribution of the Women's Health Movement has been to reveal gender bias in medical knowledge and practice (Ruzek 1978; Zimmerman 1987; Weisman 1998). Women's health activists, feminists, and women physicians of the nineteenth century fought against misogyny in medical science and medical practice in order to make this point, challenging at the same time powerful cultural ideals that limited women such as themselves to a narrow set of domestic and "private" roles (Walsh 1977). They wrote and campaigned for women's health issues, struggling against male dominance along with midwives, nurses, and a growing number of women physicians and researchers, laying the groundwork for changes that would not occur until much later (Monteiro 1984). These efforts continued into the twentieth century (Barker 1998), fluctuating along with the political climate, yet keeping the fundamental impulse of the movement alive. In the 1990s, women's health captured enough public momentum for some of its claims to be enacted into public policy, which brought funding to support women's health research, career development of women in the biomedical sciences, and reforms in the training of physicians (Weisman 1998).

The changes brought to fruition in the 1990s signaled a major shift in health care delivery for women. They underscored the importance of the Women's Health Movement in offering the first significant and broad-based critique of modern medicine and served, perhaps, as the first major consumer voice in health care for the twentieth century (Rodwin 1994; Zimmerman and Hill forthcoming). In this paper, I address this concern with the quality of health care by examining the connections between women's health, health care, and gender bias in medical education. My purpose is to comprehensively assess the current status of women's health in medical education and training by focusing on three important areas where gender bias has the potential to influence the culture and process of medical care. Firstly, I consider the presence of women as medical students and faculty, and their participation in the hierarchical structure and decision-making of U.S. medical schools. Next, I take up the issue of gender bias in the content of medical curricula and training programs. Finally, I turn to a consideration of the friendliness for both men and women of the climate and environment of medical education, arguing that if physicians are to respect and give attention to women's health they should see these issues receiving priority in medical schools. The paper concludes with a discussion of several emerging trends and challenges that affect the future status of women's health in medical education.

WOMEN'S HEALTH IN MEDICAL EDUCATION AND TRAINING

There is wide consensus that women's health requires more attention within health care education, training, and professional organizations than it currently receives (Altekruse and McDermott 1988; Donoghue 1996; Levison 1994; Lowey 1994; Wallis 1993; U.S. Department of Health and Human Services 1995a). This point of view had its recent origins some 30 years ago when early voices from the Women's Health Movement began to advocate for improved health care for women (Ruzek 1978; Weisman 1998). At that time it was thought—quite narrowly, in retrospect—that the key to change was to increase the number of women physicians. Now, even though women physicians have been found to practice differently than their male counterparts—more involved in primary care, spending more time with patients, establishing more egalitarian doctor–patient relationships, and practicing more preventive medicine (Lurie, Slater, McGovern, Ekstrum, Qual, and Margolis 1993; Mendelsohn, Nieman, Issacs, Lee, and Levison 1994; Bertakis, Helms, Callahan, Azari, and Robbins 1995)—it is recognized that simply increasing the number of women physicians is not sufficient to remedy the inadequacies of women's health care. The status of medical knowledge about women and the biases inherent in both the content and process of medical education have been identified as in need of revision. In its 1995 report, "Women and Medicine," the Council of Graduate Medical Education (COGME) recognized the centrality of these two critical and interrelated issues and called for changes both in physician education in women's health and the status of women in the medical profession (U.S. Department of Health and Human Services 1995b). Reconstituting the body of knowledge taught in medical schools and other health care training programs to include more women's health is intended to enhance the delivery of health care and, ultimately, to improve the quality of life for women.

The following discussion concentrates on women's health in relation to the education of physicians, an area of considerable attention and a growing scholarly literature. Many similar problems, deficiencies, and issues can be found in the education of other health care providers—including, pharmacists, physical and occupational therapists, health care administrators, public health specialists, dentists, and nurses—as well as throughout the sciences (Fox 1999; *MIT Faculty Newsletter* 1999) Therefore, while medicine is the specific focus here, the review and analysis of literature, the issues and trends covered, the challenges raised, and the policy initiatives summarized can be helpful to health education and training beyond medicine.

Many of the problems and issues facing women medical students, residents, and faculty are compounded when these individuals are also women of color and/ or represent ethnic minorities. The literature on minority women's health is quite limited; where it is possible, however, this paper gives special consideration to issues as they affect these women.

This review and analysis is organized around three issues. The first is the question of women's access to medical education and to educational leadership positions within medical schools and national medical education organizations (i.e., their recruitment, retention, and promotion). The second focuses on women's health content and the organizational format of the medical curriculum. The third issue concerns the quality and atmosphere of the learning environment, especially the establishment and maintenance of a respectful and supportive environment for women students, residents, and faculty.

WOMEN'S ACCESS TO MEDICAL EDUCATION AS STUDENTS AND FACULTY

The equitable representation of women in medical schools as students, residents, researchers and faculty is an important component in the preparation of physicians to meet the health needs of women. Gender balance in medical schools helps assure that women will have (1) a supportive cultural environment that addresses the needs of both male and female students and faculty; (2) the availability of women faculty as role models and mentors both in clinical medicine and in research; (3) auxiliary academic programs and a core curriculum that are balanced in terms of covering gender and sex differences, that emphasize both male and female models of the human body, and that are respectful to both men and women; and (4) expanded opportunities for women to move into leadership roles where they can assume more responsibility in shaping the future of U.S. medicine and health care.

Proportional Representation of Women and Minorities

In the 1998 to 1999 academic year, 44.4% of entering medical students and 42.6% of U.S. medical school graduates were women (Barzansky, Jonas, and Etzel 1999). This represented well over a threefold increase since 1970 when only 10% of entering students and 9% of graduates were women (U.S. Department of Health and Human Services 1995b). Whether or not the dramatic rise in the proportion of women in U.S. medical schools has reached a plateau is yet to be determined; however, it is noteworthy that some of the nation's most prestigious medical schools recently have had more women than men among their entering students—for example, in 1994, women composed 53% of the first year class at Johns Hopkins, Harvard, and the University of California San Francisco and 56% of the class at Yale. At the same time, other schools had proportions of women as small as 25% (Bickel, Galbraith, and Quinnie 1995).

The racial-ethnic diversity of U.S. medical schools also has increased. In the academic year from 1984 to 1985, 16% of U.S. medical students had African-American, Native-American, Hispanic, or Asian backgrounds (Barzansky, Jonas, and Etzel 1995); by 1998 to 1999, this percentage had increased to 34% (Barzansky, Jonas, and Etzel 1999). Of these, however, more than half (55%) were Asian-American students, meaning that the other minority groups were "underrepresented minorities" (i.e., represented in medicine below their proportion in the general population). For example, nationally, African-Americans represent approximately 12% of the population, but only about 3.7% of practicing physicians (Taylor, Hunt, and Temple 1990) and only about 8% of enrolled medical students in 1999 (Barzansky, Jonas, and Etzel 1999). In fact, in 1997, applications from underrepresented minorities decreased 11% from 1996 and the number of entering underrepresented minority students decreased 8.4% (Barzansky, Jonas, and Etzel 1998).

Women medical students have become more ethnically diverse than their male counterparts. Over the same fourteen years, the percentage of women among minority students more than doubled, reaching 37% in 1999. While underrepresented minorities constituted 13% among male medical students, they constituted 19% of the women. This general pattern of increase in women students has occurred despite the fact they have fewer financial resources and frequently carry higher loan balances into practice. According to the American Medical Women's Association, 79% of female medical students struggle financially compared with 58% of male students. The Association of American Medical Colleges reports that women enter and depart from medical school with greater debt loads than men (Bickel and Ruffin 1995).

Progress in gender equity is less visible in residency programs than in undergraduate medicine. In the 1998 to 1999 academic year, 37% of resident physicians were women, a smaller proportional representation compared to undergraduate medical students (Miller, Dunn, and Richter 1999). Although not discussed in the literature, this raises the question of whether women are less likely than men to move directly into residency programs. Of the 98 specialty and subspecialty residency programs listed by the American Medical Association in 1998, 94 (96%) included women residents (Miller, Dunn, and Richter 1999). Even so, women tend to "cluster" in a relatively limited number of specialty choices; about 60% of practicing women physicians are either in family practice, internal medicine, obstetrics/gynecology, pediatrics, or psychiatry. In 1999, 72% of women residents were in these specialities compared to 53% of men (Miller, Dunn, and Richter 1999). These data also show that a number of medical residencies remain significantly gender-segregated. Examples include surgery, where just 21% of general surgery residents, 7% of orthopaedic surgery residents, and 6% of thoracic surgery residents were women in 1998 compared to the proportions of women residents in family practice (46%), obstetrics/gynecology (64%), and pediatrics (64%). This clustering and uneven distribution of women raises the question of self-selection versus barriers to access for women, issues that will be considered later in relation to the gender climate in medical education. Progress in gender equity is slower outside of academic medicine, where about 20% of practicing physicians currently are women; an increase of 30% by the year 2010 is expected (American Medical Association 1995).

Increases in the numbers of women and minorities in U.S. medical schools may have a positive effect on the performance and overall well-being of students in these two groups. Sociological research suggests that, to the extent individuals in noticeably distinctive social groups are in the minority within a large social setting (e.g., women and minorities within medical schools), their visibility makes them subject to particular patterns of behavior (Kanter 1978, Dufort and Maheux 1995). For example, they may experience stereotyping and greater performance pressures without the benefit of a large social support network, leading them to react with hostility or by withdrawing. Those in the minority may be easily intimidated and uncomfortable expressing perspectives that differ from the majority. While these dynamics cannot be eliminated from social life, they can be ameliorated by greater gender and racial/cultural balance.

Career Progress and Mentorship

Research on the careers of women in medicine shows that, as women move through graduate medical education and into professional careers, role models and particularly mentor relationships are vital for advancement (Lorber 1984). With relatively few women physicians practicing in the community, the existence of women as role models and potential mentors within medical faculties is particularly important. Unfortunately, women physicians have experienced a "glass ceiling" within many medical schools, moving into positions of authority less frequently and more slowly than men. In 1995, 25% of U.S. medical school full-time faculties were women (Bickel, Galbraith, and Quinnie 1995) a proportion substantially lower than the proportion of women medical students. This trend was even more pronounced when the relationship between the proportion of minority women medical students (37%) is considered in relation to the proportion of minority women medical faculty (17%) (Palepu et al. 1998). More problematic, however, was that the ratio of women to men became even smaller among upper faculty ranks, those positions in which women could most effectively mentor students. While 32% of male faculty members were full professors, only 10% of female faculty held that rank. The fact that these figures changed very little between 1991 and 1980, when 9% of women faculty and 30% of men were full professors (Wear 1994), has led to interest in studying the causal factors, including the possibility of bias in the promotions process.

There are no clear explanations for women's attenuated career progress. Kaplan and her colleagues (1996) conducted a national study of academic advancement among pediatricians and concluded that sex differences in academic rank (but not salary) could be explained by a complex of factors, including greater time spent in teaching and patient care, less scholarly productivity, and less perceived institutional support. In this study, 23% of highly productive women perceived barriers to advancement compared to only 7% of highly productive men (Kaplan et al. 1996). Bickel (1995) has suggested a process of "cumulative disadvantage" that includes sexist practices (e.g., collegial exclusion and harassment under the guise of joking), the inflexibility of organizational structures, and the lack of mentoring. Upper level faculties have an important role in the decisions that shape medical education and are particularly influential in affecting the nature of the medical school environment and culture. Only a small proportion of medical school departments (approximately 2%) are headed by women (Mandelbaum-Schmid 1992). Seven women and 118 men held allopathic medical school Dean positions in 1998 according to the AAMC, an increase of three since 1991. Between 1991 and 1994, female associate deans increased from 114 to 131 and assistant deans from 102 to 118. As evidenced by the gender-related programs of the AMA, AAMC, and other professional organizations such as American Medical Women's Association (AMWA), as well as by activities in various medical schools around the country, increased awareness of gender-related issues and improved conditions for women in medical schools have coincided with the increasing proportion of women medical students and faculty. Clearly, however, the full effect of numbers is limited by the barriers and difficult choices women face as they move into more powerful positions.

WOMEN'S HEALTH IN THE MEDICAL CURRICULUM: CONTENT AND STRUCTURE

As a result of grass roots pressure and the legislative response to it, medical education has come under scrutiny over the past several years in order to examine whether physicians are receiving appropriate education and training to meet the needs of women patients. The very basis of medical knowledge is at issue: whether or not the research and clinical information used to make preventive, diagnostic, and treatment decisions about women is adequately comprehensive and has been derived from studies of women rather than simply generalized from knowledge of men. There is little doubt that medical research on women's health issues and the use of women subjects have been neglected and that competence in cultural diversity and in women's health is in need of strengthening. Both the content of knowledge and the way knowledge is organized and transmitted are important in assessing medical education's treatment of women's health.

Curricular Content

Deficiencies have been identified in the knowledge content of medical school, residency, and continuing medical education curricula stemming from (1) the inadequacy of medical research on women (Kirschstein 1991; Dresser 1992; Rosser 1994), (2) gaps in the extent to which existing research findings on women are incorporated into medical curricula (Roberts, Kroboth, and Bernier 1995), and (3) the observation that medicine has developed on a model of the male body, with women conceptualized as "the other," a deviation from the male norm (Harrison 1990). Deficiencies may also exist in the presentation of women's normal development and functioning, and the extent to which it is similar to or distinct from men's. For example, according to Levison (1994), problems in insulin management in some diabetic women that a physician might attribute to poor adherence to diet and insulin administration may, in fact, occur because changes in insulin requirements are related to effects of hormone fluctuation associated with the menstrual cycle.

While much current criticism focuses on the gaps and lack of specific attention to women's health, medical curricula have also been faulted for maintaining a historically outdated preoccupation with women's reproductive systems—that is, an abundance of attention on reproductive systems so that a comprehensive approach cannot be achieved. Furthermore, medical knowledge and medical education have, in past instances, incorporated cultural assumptions and attitudes and values that devalue women, reflect limited, subordinate roles for them, and perpetuate negative stereotypes (Scully and Bart 1973; Federation of Feminist Women's Health Centers 1981; Giacomini, Rozee-Koker and Pepitone-Arreola-Rockwell 1986; Zimmerman 1987; Mendelsohn et al. 1994). It is important to note that the male norm in medicine is actually a Caucasian, middle-class male norm, and that the lack of attention to women is paralleled by a lack of attention to people of color and to those with diverse sociocultural characteristics (Roberts, Kroboth, and Bernier 1995).

As a consequence of these problems, medical curricula have not adequately integrated knowledge of sex differences in diseases and conditions that occur in both men and women. In addition, medical curricula often ignore or give too little attention to diseases and conditions that are unique to women or that are more prevalent or more serious among women. Similarly, inadequate attention may be given to risk factors that, for some diseases or conditions, may be different for women than for men or to certain diseases or conditions where the most effective interventions may differ by gender. Similar gaps exist in incorporating knowledge about racial and ethnic minorities. These gaps are of major concern because women may incur significant health risks as a result—risks compounded by the fact that women receive medical care more often than men in the form of more physician contacts, prescriptions, and hospitalizations (Zimmerman and Hill 1999).

The standard medical curriculum also includes instruction in the process of clinical care and doctor–patient communication. Male physicians have been criticized frequently for the way they respond to women patients, which is not entirely surprising given that gender and women's health competencies are usually only a small part, if any, of their education. There is evidence that male physicians often doubt the credibility or the accuracy of women's accounts, fail to acknowledge, respect, or show sensitivity to women's sociocultural position and obligations and, in other ways, exhibit attitudes and behavior that devalue women (Fisher 1986; Miles 1991).

One of the difficulties in attempting to alter medical education is that much clinical teaching is done on an apprenticeship or tutorial basis where students learn by emulating their teachers. Thus, teachers must change before their students can. Because physicians so often are teaching students by example, teacher awareness of and sensitivity to gender bias is very important. In addition to improving curricular content on the etiology, diagnosis, treatment, and prevention of diseases in women, sensitivity and doctor-patient communication can be enhanced through placing greater educational attention on sociocultural, economic, and behavioral issues, including gender issues. If physicians understood more about issues such as the dynamics of poverty, violence, and economic dependence, women's dual responsibilities at home and at work, and the cultural meanings associated with gender roles, they might more effectively respond and interact with women patients as well as colleagues and students. There is evidence from pilot programs that men as well as women students would be receptive to greater emphasis on women's issues and gender roles in medical curricula (Hohener and Spielvogel 1995). In the mid-1990s, The National Academy on Women's Health Medical Education created women's health competencies for residency training in internal medicine that added to similar competency expansions in family practice and obstetrics/gynecology (Donoghue 1996). These curriculum reform efforts in medicine have been confounded by the structural fragmentation of health care for women in multiple medical specialties. In other health fields, such as Pharmacy and Physical Therapy, curriculum reform efforts are just beginning.

Curricular Structure

In addition to reforming the content of medical education, the format or structure of the curriculum has been of concern. The major problem here is that women's health is fragmented, partitioned according to specialization and professional turf interests rather then women's health care needs (Hoffman and Johnson 1995). Fragmentation is a particular problem for women because routine care for their common health problems is divided in the organization of medical education and medical practice between gynecological/obstetrical and other health services. Further fragmentation stems from the existence of multiple providers for many of these two types of services: routine care for common reproductive and nonreproductive problems can be provided by either family practice, internal medicine or obstetrics and gynecology (Ob/Gyn). All three specialties see patients for many of the same problems, resulting in overlap with the potential for duplication or gaps in care, and problems with coordination and continuity of care for patients. Data from the National Ambulatory Medical Care Survey show, for example, that women see all three specialities for the following:

- *Upper respiratory disorders*—72% consult family practice, 24% consult internal medicine, and 3% consult ob/gyn
- *General medical exams*—30% consult family practice, 13% consult internal medicine, and 57% consult Ob/Gyn
- *Urinary tract infections*—60% consult family practice, 23% consult internal medicine, and 17% consult Ob/Gyn
- *Gynecologic disorders*—28% consult family practice, 4% consult internal medicine, and 67% Ob/Gyn(Bartman and Weiss 1993)

While there is a general consensus about the content that should be included in medical curricula—the Council on Graduate Medical Education, for example, has underscored that all physicians should have competency in women's health (U.S. Department of Health and Human Services 1995b)—there are differences of opinion over the way women's health should be configured. Specifically, there has been a debate over whether women's health should be a separate speciality or subspeciality, or whether, on the other hand, it should be integrated and taught to all students throughout their medical education. Furthermore, there are disagreements and political concerns over which of the existing specializations should take primary responsibility for the care of women.

During the last several decades, routine health care for many women has been split between obstetrics/gynecology and internal medicine. For adequate care, women have required two physicians. It is unknown how many women actually have had this form of care; however, today this division of labor is particularly problematic when many women are uninsured or limited under managed care plans to a single primary care physician. A third specialty, family practice, also is prominent as a provider of health care for women. This multiple-provider situation is at the root of the debate over how and where to incorporate women's health within the medical curriculum. Who should provide routine health care for women? Who should be trained in women's health? Dr. Michelle Harrison (1990) has summarized it this way: "The student planning to 'do women's health' then must be either an internist who learns obstetrics–gynecology elsewhere, an obstetrician-gynecologist who learns medicine elsewhere, or a family physician—in which case the price of learning to take care of a woman is learning to take care of the rest of her family also." All three specialities share the care of women patients, yet each has a different program of training, and none as yet can specifically train physicians to deliver comprehensive care to women based on information derived from research and clinical training with women.

Obstetrics and gynecology has traditionally been perceived as the women's health specialty, reflecting a (now largely outmoded) view of women's health as primarily reproductive care. Views within contemporary medicine are divided on the appropriate definition and location of Ob/Gyn in the spectrum of care. Opinions range from those who see it as largely a surgical specialty to others for whom it is women's primary care. Education in Ob/Gyn, until recently, has not incorporated much comprehensive training on nonreproductive aspects of women's health; however, in 1995, the Council for Residency Education for Ob/Gyn redesigned its curriculum to provide broader coverage of women's health. More education on common chronic conditions women face, such as diabetes and hypertension, and also greater attention to basic primary care, including coverage of psychosocial issues, were among the changes.

Internal medicine historically has focused on women's health as a distinctive body of knowledge mainly in the area of the reproductive system, using the generic male model for other bodily systems and functions. Internists today may do little or no reproductive health care, referring patients to Ob/Gyn physicians instead. They receive little training in social and behavioral issues of special concern to women, such as depression, poverty, economic dependence, violence, incest, eating disorders and substance abuse. The American Board of Internal Medicine in 1995 published a position paper on women's health, recommending that

1. Women's health should not be separated from the larger role of the generalist physician.
2. Recognizing women's health specialists by certification risks fragmenting the profession.
3. Women's health is multi disciplinary.
4. General internal medicine should facilitate continued growth in women's health education and research. (Donoghue 1996)

Family practice is the nearest among the three to providing broad education and training that includes psychological and social perspectives; yet it, too, is based largely on the standard, male model. In 1994, the American Academy of Family Physicians issued recommended core educational guidelines for Family Practice residents in the area of Women's Health. The guidelines included recommendations for physician attitudes toward women as well as women's health knowledge and skills.

In addition to changes in these speciality fields, there have also been recent women's health initiatives within traditional medical organizations such as the Division of Women's Health Issues within the American Medical Association and the American College of Obstetrics and Gynecologists (1995) to complement the activities of the American Medical Women's Association (Donoghue 1996; Gonzalez-Pardo 1993). There are also increasing number of residency and fellowship programs in women's health (U.S. Department of Health and Human Services 1996).

GENDER FAIRNESS AND THE MEDICAL EDUCATION ENVIRONMENT

Promotion of women's health in medical education also depends on the gender climate and atmosphere associated with the educational process. Students become educated and socialized according to the way gender issues are handled within the environments of their own medical institutions, and they also may be personally encouraged or discouraged in their own medical careers as a result. The recent report "Women in Medicine," by the Council On Graduate Medical Education (U.S. Department of Health and Human Services 1995b) concluded

> Despite their increasing numbers, women physicians continue to be subjected to overt and sub-tle gender bias in professional interactions and opportunities regardless of their field of prac-tice, extent of training, professional accomplishments, and professional goals. *Gender bias is the single greatest deterrent to women physicians achieving their full potential in every area and aspect of the medical profession and across all stages of medical careers.* (italics mine)

The history of women in medicine is a record of achievement. At the same time, it presents a chronology of resistance and overt hostility toward women physi-cians and their professional advancement (Walsh 1977). Despite evidence of con-tinuing progress, there remain indications that the atmosphere in U.S. medical schools may be less friendly to women than to men (Ehrhart and Sandler 1990; Bickel and Ruffin 1995). In research on medical students and residents, for exam-ple, 42% of women (compared to 36% of men) reported being belittled or humil-iated in public, and between 24 and 80 percent of the women surveyed reported experiencing various forms of sexual harassment (Komaromy, Bindman, Haver, and Sande 1993). The climate of medical education may include patterns that are often difficult to articulate or measure, such as patronizing or condescending atti-tudes toward women, subtle preferences in the treatment of men, lack of openness in collegial opportunities for women faculty, inadequate representation of women as visiting professors and guest lecturers, sexist pictures or other materials used in classes or visible in other settings, and a paucity of resources that address women's issues. Evidence indicates that these features of the environment are clearly visible and often problematic for women, while they often go unnoticed or are viewed as relatively minor issues by men (Hostler and Gressard 1993). Gen-der bias can be powerful enough to skew perceptions—albeit unknowingly—cre-ating a culture than devalues women. This also occurs in other academic settings as suggested by a recent study of 147 chairs of Psychology Departments in which fictitious researchers' credentials were consistently evaluated at the lower rank of assistant professor when identified as female, and to the higher associate professor rank when identified as male (Alexander 1989).

Women, who are members of minorities, are potentially at double jeopardy if the medical education climate also involves racial or ethnic bias. A recent study indicated that 32% of white students reported they were victims of racial or ethnic slurs compared to 39% among under represented minorities (Baldwin, Daugherty, and Rowley 1994).

The degree to which tension and conflict between medical careers and family is officially recognized also indicates an institutional climate friendly to women (although both male and female students benefit from family friendly policies). The vast majority of physicians have both spouses and children. Balancing career work and family work, while infrequently raised by men, typically is high on the list of concerns of medical women, both students and faculty. The expectations women face for unpaid "second shift" work in the household and as family care-givers are more extensive than those faced by men. This considerable difference in domestic and caring responsibilities and expectations in the lives of medical men and women creates the conditions for barriers to understanding and communication as well as for discrimination and gender bias. Even before admission to medical school, women applicants report being questioned more intensely than men about their marriage and family plans (Grant 1988). The lack of institutional awareness and support for women's family roles is evidenced in the medical environment both by the lack of interpersonal sensitivity and by inadequate policies to assist women during childbearing and parenthood (Philibert and Bickel 1995; Levinson, Tolle, and Lewis 1989; U.S. Department of Health and Human Services 1995c). Policies in need of improvement include insurance coverage, maternity leave, and parental leave opportunities, childcare, and flextime work arrangements. Medical school policies for part-time faculty committed to "full professional effort" (i.e., FPE faculty) are also a component of a women-friendly institutional climate. A recent survey of FPE policies found that 45% of U.S. and Canadian medical schools reported having specific procedures in place (Froom and Bickel 1996). Whether in response to the difficulties of combining family and a career in medicine or for other reasons, AMA data show that women physicians are less likely to marry (86% versus 95%) and to have children (85% versus 93%) than their male counterparts.

Married women physicians often find themselves involved in stressful dual career families. They tend to marry other professionals (70% of married women physicians are married to professionals) and half marry other physicians. A notable gender difference that may well affect understanding and communication between male and female physicians is the fact that, while 88% of women physicians have spouses who are employed full-time, only 26% of men physicians do (Jackson 1993). Thus, while a woman physician is likely to be struggling with dual responsibilities, a man is much more likely to be married to a full-time homemaker. The realities of marriage and family responsibilities for women physicians are reflected in the small but consistently found gender difference in the number of hours worked per week (women working slightly less)(Lee and Mroz 1991). This finding has been shown to disappear among women who have no children, suggesting that it is parenthood that affects work hours and only among mothers, not among fathers (Grant, Simpson, and Rong 1990). There is reason to assume that these findings may also reflect the stressful situations of medical students and residents with husbands and children. Supportive educational environments would be those that facilitated maternity and parenting through programs such as flexible scheduling, leave programs, shared residency options, and convenient childcare.

When these various components of the medical education climate are considered, the evidence suggests that, though improvements are underway, gender fairness has not yet been achieved. Even when curricular changes are made, it is difficult to imagine how women's health can be effectively communicated if the medical school climate contradicts the very cultural and gender competencies being taught. As Dr. Frances Conley (1993) observed after resigning from a prestigious medical faculty position in protest over years of gender bias, "technical talent and academic credentials must be coupled with decent behavior." The cultures of medical schools, though difficult to see or control, are nonetheless critical to the successful integration of women's health into the curriculum and training process of physicians.

EMERGING TRENDS AND CHALLENGES FOR WOMEN'S HEALTH IN MEDICINE

It is too early to know if the women's health policy changes that took place in the 1990s were part of a "paradigm shift" in the way women's health care is delivered, or whether they were more routine, episodic events—a "wave" in the women's health "megamovement (Weisman 1998)." We can, however, draw some conclusions about current trends and the problems and issues they raise for the future of women's health in medicine.

Women, during the twentieth century, became an integral part of medical education as students, faculty, and as administrators. Nonetheless, numerous factors still preclude the full entry of women into positions of leadership, decision authority, and effective mentoring, within U.S. medical schools as well as within major professional organizations (Tesch, Wood, Helwig, and Nattinger 1995). Data show, for example, that changes have occurred more slowly at the higher ranks where academic promotions have not been consistent with women's qualifications and achievements. Where programs to promote women's advancement in medicine have been developed, they typically have focused on improving the characteristics and qualifications of individual women (Bickel and Whiting 1991). Programs have neglected the structures and policies of the organizations themselves. Johns Hopkins University, however, provides an exception. When an internal study of the status of women in its medical school was conducted in 1987, officials responded with not only individualistic solutions, but also with institutional changes such as a revised faculty review process, changes in the promotions process, and changes in the way school committees were constituted (DeAngelis and Johns 1995).

The curriculum in medicine is being reevaluated and redesigned so that it better incorporates subject matter on women's health. Much of the prior inadequacy stemmed from the fact that not enough research has been conducted specifically on the health problems of women and/or using women as research subjects. Furthermore, curricula have failed to adequately incorporate existing knowledge about women's bodies, development, and gender-specific aspects of diseases. Curricula have also failed to integrate sociocultural, economic, and behavioral perspectives along with the biomedical. The future challenge is for medical education to continue the gender reform process begun in the 1990s until balance and equity are achieved. The gender climate in medical education has also begun to receive attention. The increasing diversity among medical students—larger proportions of women and increasing proportions of minorities—requires cooperation and understanding among individuals with quite different cultural backgrounds and perspectives. This suggests that cultural and gender competencies will become an increasingly important part of medical education.

The future of women's health is tied to changes broader than the three areas I have discussed in this paper. The private market economy of American health care is in the process of appropriating more and more physician autonomy, bringing into question the ability of medicine to bring about significant changes in the structure of medical knowledge and practice from within. Health insurance in the United States, whether public or private, has been found to serve men's needs better than women's (Miles and Parker 1997). Managed care holds the promise of improved and more efficient care; however, consistent with the corporatization of medicine, financial outcomes under managed care appear to be taking priority over medical ones. While there are increases in screening services for women, infertility treatment and mental health services are often inadequate (Zimmerman and Hill forthcoming). Managed care limitations on mental health–related hospital stays, for example, make it difficult for anorexia nervosa patients, mostly women, to receive appropriate treatment. Even though physicians are emerging from medical schools equipped to provide women better medical care, the gender bias in insurance coverage can thwart their efforts.

The tough challenges for the twenty-first century revolve around the extent to which U.S. medicine—now that it has finally begun to direct itself toward gender, balancing the curriculum and improving its coverage of women's health in medical education and training—will, in fact, be able to exert the leadership necessary to accomplish this objective.

REFERENCES

Alexander, N. 1989. "The Glass Ceiling: Science and Medicine," *Fertility News* 23:10.

Altekruse, J.M. and S.W. McDermott. 1988. "Contemporary Concerns of Women in Medicine." Pp. 65-88 in *Feminism within the Science and Health Care Professions: Overcoming Resistance*, edited by S.V. Rosser. New York: Pergamon Press.

American Medical Association. (AMA). 1995. *Diagnostic and Treatment Guidelines on Family Violence*. Chicago: AMA Department of Mental Health.

Baldwin, D.C., S.. Daugherty, and B.D. Rowley. 1994. "Racial and Ethnic Discrimination During Residency: Results of a National Survey." *Academic Medicine* 69: S19-S20.

Barker, K. 1998. "Women Physicians and the Gendered System of the Professions," *Work and Occupations* 25(2): 229-255.

Bartman, B.A. and K.B. Weiss. 1993. "Women's Primary Care in the United States: A Study of Practice Variation Among Physician Specialties." *Journal of Women's Health*. 2: 261-268.

Barzansky, B., H.S. Jonas, and S.I. Etzel. 1995. "Educational Programs in U.S. Medical Schools, 1994-1995." *Journal of the American Medical Association* 274: 716-722.

———. 1998. "Educational Programs in U.S. Medical Schools, 1997-1998." *Journal of the American Medical Association* 280(9): 803-806.

———. 1999. "Educational Programs in U.S. Medical Schools, 1998-1999." *Journal of the American Medical Association* 282(9): 840-846.

Bertakis, K.D., L.J. Helms, E.J. Callahan, R. Azari, and J.A. Robbins. 1995. "The Influence of Gender of Physician Practice Style," *Medical Care* 33(4): 407-416.

Bickel, J. 1995. "Scenarios for Success–Enhancing Women Physicians' Professional Advancement." *Western Journal of Medicine* 162: 165-169.

Bickel, J. and A. Ruffin. 1995. "Gender-associated Differences in Matriculating and Graduating Medical Students." *Academic Medicine* 70: 552-559.

Bickel, J. and B.E. Whiting. 1991. "AAMC Data Report: Comparing the Representation and Promotion of Men and Women Faculty at U.S. Medical Schools." *Academic Medicine* 66: 497.

Bickel, J., A. Galbraith and R. Quinnie. 1995. *Women in U.S. Academic Medicine Statistics 1995*. Washington, D.C.: American Association of Medical Colleges.

Conley, F. 1993. "Toward a More Perfect World—Eliminating Sexual Discrimination in Academic Medicine." *New England Journal of Medicine* 328: 352.

DeAngelis, C.D. and M. E. Johns. 1995. "Promotion of Women in Academic Medicine: Shatter the Ceilings, Polish the Floors." *Journal of the American Medical Association* 273: 1056-1057.

Donoghue, G.D. (Ed.) 1996. *Resource Guide for Faculty: Women's Health in the Curriculum: Undergraduate, Residency and Continuing Education*. Philadelphia: National Academy on Women's Health Medical Education.

Dresser, R. 1992. "Wanted: Single, White Male for Medical Research." *Hastings Center Report*. January-February: 24-29.

Dufort, F., and B. Mahuex. 1995. "When Female Medical Students Are the Majority: Do Numbers Really Make a Difference?" *Journal of the American Women's Medical Association* 50: 4-6.

Ehrhart, J.K. and B. Sandler. 1990. *Rx for Success: Improving the Climate for Women in Medical Schools and Teaching Hospitals*. Washington, D.C.: Project of the Status and Education of Women, Association of American Medical Colleges.

Federation of Feminist Women's Health Centers. 1981. *A New View of a Woman's Body*. New York: Simon & Schuster.

Fisher, S. 1986. *In the Patient's Best Interest: Women and the Politics of Medical Decisions*. New Brunswick, NJ: Rutgers University Press.

Fox, M.F. 1999. "Gender, Hierarchy and Science" in *Handbook of the Sociology of Gender*, edited by J.S. Chafetz. New York: Kluwer Academic/Plenum Publishers.

Froom, J.D., and J. Bickel. 1996. "Medical School Policies for Part-time Faculty Committed to Full Professional Effort," *Academic Medicine* 71: 91-96.

Giacomini, M. Rozee-Koker, P., and Pepitone-Arreola-Rockwell, F. 1986. "Gender Bias in Human Anatomy Textbook Illustrations." *Psychology of Women Quarterly* 10: 413.

Gonzalez-Pardo, L. 1993. "The Need for a Curriculum on Women's Health," Panel on Women and Medicine, Washington, D.C.: Council on Graduate Medical Education.

Grant, L. 1988. "The Gender Climate of Medical School: Perspectives of Women and Men Students." *Journal of the American Medical Women's Association.* 43:109-119.

Grant, L., L.A. Simpson, and X.L. Rong. 1990. "Gender, Parenthood and Work Hours of Physicians," *Journal of Marriage and the Family* 52: 39-59.

Harrison, M. 1990. "Woman as Other: The Premise of Medicine." *Journal of the American Medical Women's Association.* 45: 225-226.

Hoffman, E. and K. Johnson. 1995. "Women's Health and Managed Care: Implications for the Training of Primary Care Physicians." *Journal of the American Medical Women's Association.* 50: 17-19.

Hohener, H.C. and A. M. Spielvogel. 1995. "Teaching Women's Issues in Psychiatric Residency: Residents' Attitudes." *Journal of the American Medical Women's Association.* 50: 14-16.

Hostler, S.L. and R.P. Gressard. 1993. "Perceptions of the Gender Fairness of the Medical Education Environment." *Journal of the American Medical Women's Association.* 48: 51-54.

Jackson, A. 1993. "Women Physicians Practice in Less Prestigious Specialities than Men, *University of Michigan LSA Magazine,* Fall, Vol. 4.

Kanter, R.M. 1978. *Men and Women of the Corporation.* New York: Basic Books.

Kaplan, S.H., L.M. Sullivan, K.A. Dukes, C.F. Phillips, R.P. Kelch, and J.G. Schaller. 1996. "Sex Differences in Academic Advancement: Results of a National Study of Pediatricians." *The New England Journal of Medicine.* 335 (17): 1282-1289.

Kirschstein, R.L. 1991. "Research on Women's Health," *American Journal of Public Health.* 81: 291-293.

Komaromy, M., A.B. Bindman, R.J. Haver, and M.A. Sande. 1993. "Sexual Harassment in Medical Training." *New England Journal of Medicine.* 328: 322-326.

Lee, R.H., and T.A. Mroz. 1991. "Family Structure and Physicians' Hours in Large, Multispeciality Groups," *Inquiry* 28: 366-374.

Levison, S.P. 1994. "Teaching Women's Health: Where Do We Stand? Where Do We Go From Here?" *Journal of Women's Health.* 3: 387-396.

Levinson, W., S.W. Tolle, and C. Lewis. 1989. "Women in Academic Medicine: Combining Career and Family." 321: 1511.

Lorber, J. 1984. *Women Physicians.* London: Tavistock.

Lowey, N.M. 1994. "Women's Health and Medical School Curricula." *Academic Medicine.* 69: 280-281.

Lurie, N., J. Slater, P. McGovern, J. Ekstrum, L. Qual, and K. Margolis. 1993. "Preventive Care for Women: Does the Sex of the Physician Matter?" *The New England Journal of Medicine.* 329: 478-482.

Mandelbaum-Schmid, J. 1992. "An Unequal Past, A Common Future: Will Medicine Change as More Women Enter the Profession?" *MD* May: 87-96.

Mendelsohn, K.D., L.Z. Nieman, K. Issacs, S. Lee, and P. Levison. 1994. "Sex and Gender Bias in Anatomy and Physicial Diagnosis Text Illustrations," *Journal of the American Medical Association.* 272: 1267-1270.

Miles, A. 1991. *Women, Health and Medicine.* London: Open University Press.

Miles, S. and K. Parker. 1997. "Men, Women and Health Insurance." *The New England Journal of Medicine.* 336: 218-221.

Miller, R.S., M.R. Dunn, and T.H. Richter. 1999. "Graduate Medical Education, 1998-1999: A Closer Look," *Journal of the American Medical Association* 282 (9): 855-860.

MIT Faculty Newsletter. 1999. "A Study of the Status of Women Faculty in Science at MIT" XI (4): March. Cambridge, Massachusetts.

Monteiro, L.A. 1984. "On Separate Roads: Florence Nightengale and Elizabeth Blackwell." *Signs* 9 (3):

Philibert, I. and J. Bickel. 1995. "Maternity and Parental Leave Policies at COTH Hospitals: An Update." *Academic Medicine* 70: 1055-1058.

Roberts, M.M., F.J. Kroboth and G.M. Bernier. 1995. "The Women's Health Track: A Model for Training Internal Medicine Residents." *Journal of Women's Health* 4: 313-318.

Rodwin, M.A. 1994. "Patient Accountability and Quality of Care: Lessons from Medical Consumerism and the Patients' Rights, Women's Health and Disability Rights Movements." *American Journal of Law and Medicine*. 20: 147-167.

Rosser, S.V. 1994. "Gender Bias in Clinical Research: The Difference It Makes." Pp. 253-265 in *Reframing Women's Health*, edited by A. Dan. Thousand Oaks, CA: Sage Publications.

Ruzek, S.B. 1978. *The Women's Health Movement: Feminist Alternatives to Medical Control*. New York: Praeger.

Scully, D. and P. Bart. 1973. "A Funny Thing Happened on the Way to the Orifice: Women in Gynecology Textbooks." *American Journal of Sociology* 78: 1045.

Taylor, R.E., J.C. Hunt, and P.B. Temple. 1990. "Recruiting Black Medical Students: A Decade of Effort." *Academic Medicine* 65: 279-287.

Tesch, B.J., H.M. Wood, A.L. Helwig, and A.B. Nattinger. 1995. "Promotion of Women Physicians in Academic Medicine: Glass Ceiling or Sticky Floor?" *Journal of the American Medical Association* 273: 1022-1025.

U.S. Department of Health and Human Services. 1995a. "National Conference on Cultural Competence and Women's Health Curricula in Medical Education." Washington, D.C.: USDHHS Office of Minority Health and Office of Women's Health. October 26-28.

_____. 1995b. *Fifth Report: Women & Medicine*. Washington, D.C.: USDHHS Council on Graduate Medical Education. Publication # HRSA-P-DM-95-1. July.

_____. 1995c. "Women in Biomedical Careers: Dynamics of Change. Strategies for the 21st Century." Summary Report of the Workshop help June 11-12, 1992. Bethesda, MD: National Institutes of Health, Office of Research on Women's Health. NIH Publication # 95-3565A.

_____. 1996. *Directory of Residencies and Fellowships in Women's Health*. Washington, D.C.: USDHHS, Public Health Service, Office on Women's Health.

Wallis, L.A. 1993. "Why A Curriculum in Women's Health?" *Journal of Women's Health*. 2: 55-60.

Walsh, M.R. 1977. *Doctors Wanted: No Women Need Apply: Sexual Barriers in the Medical Profession, 1935-1975*. New Haven, CT: Yale University Press.

Wear, D. "Feminist in Medical Education: Problems and Promises." 1994. *Journal of the American Medical Women's Association*. 49: 43-47.

Weisman, C.S. 1998. *Women's Health Care: Activist Traditions and Institutional Change*. Baltimore: The Johns Hopkins University Press.

Zimmerman, M.K. 1987. "The Women's Health Movement: A Critique of Medical Enterprise and the Position of Women." Pp. 442-472 in *Analyzing Gender: A Handbook of Social Science Research*, edited by Beth B. Hess and Myra Marx Ferree. Thousand Oaks, CA: Sage Publications.

Zimmerman, M.K. and S.A. Hill. 1999. "Health Care as a Gendered System." In *Handbook of the Sociology of Gender*, edited by Janet S. Chafetz. New York: Kluwer Academic/ Plenum Publishing.

_____. Forthcoming. "Re-forming Gendered Health Care: An Assessment of Change." *International Journal of Health Services*.

INCORPORATING EMPOWERMENT INTO MODELS OF CARE:
STRATEGIES FROM FEMINIST WOMEN'S HEALTH CENTERS

Jan E. Thomas

ABSTRACT

In recent decades, consumers and providers have become increasingly frustrated with the health care system. Concerns with the "bottom line" have replaced concerns about patients and providers. In response, new models of patient-centered care being developed are encouraging participatory relationships. Patients are empowered when their values, preferences, and needs become the center of health care decisions. In this article, I illustrate how empowerment has been translated into health care settings. The data come from a qualitative study of 14 feminist women's health centers, which were established in the 1970s and had more than 20 years experience providing community services. I discuss three primary strategies they used for empowering patients: (1) education and information to increase knowledge and demystify medical procedures, (2) breaking down institutional barriers between providers and clients, and (3) providing an environment of dignity and respect. These strategies are particularly well-suited for community-based health care. This study

Research in the Sociology of Health Care, Volume 17, pages 139-152.
Copyright © 2000 by JAI Press Inc.
All rights of reproduction in any form reserved.
ISBN: 0-7623-0644-0

also contributes to the dialogue on defining "quality" of care and strategies for rethinking the dominant, medical model of health care delivery in this country.

INTRODUCTION

Recent trends in the United States' mainstream (allopathic) medical system have led to increasing dissatisfaction and confusion among consumers and providers. In the allopathic model, the institutionalized role-set between patient and provider assumes compliance and dependency in the patient and authority, objectivity, and technological expertise in providers (Parsons 1951; Ruzek 1978, 1986; Weitz 1996). It is also a model that places the patient in the position of being the object of diagnosis and treatment rather than the subject, thus ignoring the social context of patients' lives. As insurance companies and corporate offices drive more and more health care decisions, patient needs have become even more peripheral and doctors who take time to involve patients actively in decision-making are considered "unproductive" (Woolhandler and Himmelstein 1996, p. 1700). Concerns about the "bottom-line" have replaced concerns about patients. The doctor-patient relationship has become primarily a financial relationship rather than one of caring.

CONSUMERS, COST, AND MODELS OF CARE

Since the 1980s, economic, political, and social trends have set the institution of medicine on a collision course with health care consumers. An increase in consumer activism beginning in the 1960s challenged the medical system to provide a different model of care in which physicians would become "providers" or "consultants" and patients would become "clients." An emphasis on primary, preventive health care characterized this patient-centered model with the provider sharing information, discussing options, and encouraging patients/clients to take responsibility for their own care. The women's health movement was a significant factor in the development of this cultural shift toward increased consumer activism. In the 1970s, women's health activists worked to change the medical system by educating and organizing women as health care consumers.

A second trend, in direct conflict with the increasing demands of consumers, was the increasing control of the government, insurance agencies, and other third-party payer systems. The goal of this economic trend was cost containment and more "cost-effective" health care delivery. However, the results have been increased bureaucracy and complexity, an allegiance to shareholders rather than patients, and pressure on providers to decrease the length of patient visits and increase volume (Woolhandler and Himmelstein 1996; Silver 1997). These two conflicting trends—consumerism and "cost-effectiveness"—have come together in research on models of care designed to satisfy both. The patient-centered paradigm (Lathrop 1993; Gerteis, Edgman-Levitan, Daley, and Delbanco 1993; Rodwin 1994) and participatory decision making (Kaplan, Greenfield, Gandek, Rogers, and Ware 1996) are two examples of such models.

In an economic and political climate in which health maintenance organizations (HMOs) and managed care programs are pushing for shorter visits and increased volume, consumers are growing increasingly dissatisfied. Studies cited by Woolhandler and Himmelstein (1996, p. 1699) indicate patients/clients want providers to offer "more time, more information, more caring, and more mutuality in decision-making, therapy, and prevention." In addition, "HMO enrollees are more than twice as likely as fee-for-service patients to complain that care is not appropriate, that examinations are not thorough, and that physicians do not care enough or spend enough time" (p. 1700). Indicators of "quality" used by managed care organizations generally focus on cost rather than indicators associated with patient-centered care.

Kaplan and colleagues (1996, p. 497) cited several studies confirming that patients who ask questions, elicit treatment options, express opinions, and state preferences about treatments have better health outcomes and are more likely to follow through on treatment decisions than those who do not participate. In addition, when patients participate in decision-making there is an increase in patient satisfaction and loyalty to providers, and a decrease in the likelihood of pursuing litigation as a response to grievances (Kaplan et al. 1996; Hart 1992). Participatory decision making gives patients choices among treatment options and gives them a sense of control and responsibility for care (Kaplan et al. 1996, p. 498). In effect, this is a model of patient empowerment which brings their values, preferences, and needs to the center of the health care system.

The purpose of this paper is to illustrate the ways in which feminist women's health centers have translated an ideology of empowerment into health care organizations, and the role of empowerment in client-centered, participatory models of care. This study also contributes to the dialogue on defining "quality" of care and strategies for rethinking the dominant medical model of health care delivery in this country (Gerteis et al. 1993; Lathrop 1993; McKenzie 1994). The strategies for empowering clients presented in this paper are drawn from research I conducted at feminist women's health centers. These centers have been at the core of changing the way women in the United States experience health care. Women's health activists I interviewed cited several examples of the impact their centers (and the greater women's health movement) have had on mainstream medicine:

- the increase in information and education around women's health issues (seminars, books, articles)
- the new assumption that women *want* information
- informed consent protocols
- the presence of patient inserts in prescription drugs

- the increase in women providers
- warming speculums for gynecological exams
- talking with women about what is being done to them
- less intimidating waiting rooms, women's new assertiveness as health care consumers
- allocating more funding for women's health research
- developing hospital-based programs for women
- increased legislative efforts to ensure women's reproductive rights (Pearson and Seaman 1998; Thomas 1995)

In this paper, I will focus on another of their contributions—health care delivery systems, which change patterns of care, empowering clients rather than reproducing the passive, dependent role encouraged by the traditional medical model. These models of empowerment are particularly well suited for community-based health care as they situate the locus of control with the client and the community and seek to provide quality care with limited resources.

METHODS AND DATA ANALYSIS

Data for this study were drawn from surveys, interviews, and site visits at feminist women's health centers across the country. My original interest in these organizations was to study their strategies for longterm survival and their structural changes over time. Most of the feminist women's health centers began as collectives or had some form of participatory democracy. I was interested in how many of them had retained these structures and what types of issues directed organizational change. This research resulted in the development of three ideal types, which typified the organizational structures in the 1990s: feminist bureaucracies, participatory feminist bureaucracies, and collectivist democracies (see Thomas 1995). While some of these "feminist" organizations had become quite hierarchical, others had prioritized an internal structure of shared power (Thomas 1999). What intrigued me was that, regardless of structure, all of these centers had continued to provide services focused on empowering those they served. That observation led to the research on which I have based this paper.

I constructed a brief survey, which I used to locate centers that were established in the 1970s, were women-owned and operated, and were based in the feminist philosophy of self-help, education, and empowerment. I then used these criteria to classify organizations as "feminist" women's health centers (Foster 1991; Ruzek 1978; Zimmerman 1987). The survey was sent to 77 centers, which Morgen and Julier (1991) had identified as women's organizations because they "appeared to be either women controlled health clinics or feminist health advocacy and/or health education organizations" (p. 1). The questionnaire also asked about the type of services offered, average length of appointments, use of lay health workers and volunteers, and major reasons for the center's longevity. From the original 77 women's health organizations, only two dozen met the above criteria of "feminist" women's health centers. I selected a purposive sample of 14 from this list based on the following criteria: (1) geographical location, (2) historical significance, and (3) willingness to participate in the study. I visited 12 centers and conducted one-hour phone interviews with administrators from two other centers.

In total, I interviewed 34 health activists and employees between October 1992 and April 1993. The women I spoke with had worked at the women's health centers an average of 10.6 years (range=1 to 21 years) and many were in administrative positions. Several, in fact, had been founding mothers of the organizations. The semistructured interviews focused on the early history and philosophy of the center, changes in structure, philosophy, and services, and the role of feminist women's health centers in the larger women's health movement. The average length of each interview was three hours. These interviews were taped and transcribed. I also asked to review historical documents including: newsletters, organizational meeting minutes, articles written by the center's staff, videos, and news clippings. For this paper, I went back to the taped transcripts, center brochures, written materials, and articles about the various health centers and began trying to assess the specific strategies they used to empower their clients. While I felt I would somehow "know" empowerment if I saw it in the data, defining and operationalizing the concept proved somewhat difficult. In the next section, I will describe how empowerment has been defined and how it has been incorporated into feminist models of care.

CONCEPTUALIZING EMPOWERMENT

At the individual level, empowerment is a *process* (Merzel 1994, p. 410) through which one gathers information, makes choices, and receives support in a dignified and respectful environment. Empowerment challenges basic power relations (Bookman and Morgen 1988, p. 4) and provides a means of resisting the passive and dehumanizing role assigned to patients in the health care system. The individual becomes the subject of care (the client) and the object (the patient). The locus of control shifts from provider to client. At the organizational level, empowerment is an active rather than passive model of care. It is at the heart of participatory and patient-centered models of care. Involving patients/clients in decision making and treatment options, recognizing the social and cultural context of their lives, encouraging and supporting their autonomy and self-interest, and having providers share knowledge and coordinate care are all structural aspects of participatory models which empower patients/clients.

Empowerment has been a basic feminist strategy since the development of consciousness-raising groups in the late 1960s. Initially, empowerment helped relate the "personal" to the "political" (Guttmacher and Leeds 1994, p. 413) through sharing experiences, information, resources, and skills, and by offering support and political analysis. This model, an essential element of feminist strategy and activism, was easily adopted by the women's health movement in the 1970s and became ingrained in organizational structure. The mission statements of many feminist women's health centers illustrated the incorporation of empowerment at the organizational and individual level:

> The Richardson Women's Health Center is a *women-controlled* and operated health resource. Our services are based on a philosophy of *education* and *self-help*. The central idea of self-help is that every woman has the right to know about her body in order to *take control of her health* and *her health care decisions*. (Brochure from Richardson Women's Health Center, emphasis added)

And, the brochure of another center declares

> Our goal is to provide quality, affordable, *non-judgmental*, comprehensive health care *for women by women*. We are committed to *serving all women*: with a focus on lesbians and special outreach to women of color, low-income women, older women and differently abled women. We believe that health care must be *demystified* so that women can be *involved in their own care*. We believe that health care must be accessible and offered in a *safe and empowering environment*. We view ourselves as *advocates*, *activists* and *educators* on women's health. (emphasis added)

In both examples, empowerment takes place over time through the mutual sharing of information, knowledge, and skills. This culminates in a woman's active control of her health care.

Hospital-based women's health centers have also adopted these goals as part of their discourse. For example, the mission statement from The Women's Center, affiliated with Floyd Medical Center in Georgia, states, "Emphasis at The Women's Center is on empowering each woman to take an active role in keeping well and sustaining a healthy lifestyle" (workshop handout from the National Association of Women's Health Care Professionals annual meeting 1996). Increasing emphasis on care that is "personalized," is "comprehensive," "focuses on the whole person," and is a "shared responsibility" are other examples of how mainstream medical programs have adopted the rhetoric of empowerment. However, notably absent in these examples is the actual shifting of the locus of control from provider to patient. The patient is given information but is not necessarily the agent of control.

EMPOWERING CLIENTS

Patient-centered care has been the cornerstone of feminist women's health centers since their inception. The feminist women's health centers in this study focused their services and trained their staffs in ways that helped empower their clients and provided personalized care. The centers used three strategies to accomplish this: (1) education and information to increase knowledge and demystify medical procedures; (2) breaking down institutional barriers between providers and clients; and (3) providing an environment of dignity and respect. I will briefly illustrate how each of these strategies have been implemented in the feminist women's health centers.

Knowledge is Power: Providing Information and Demystifying Medicine

One of the most basic ways that feminist women's health centers empowered their clients was by providing them with education and information in a nonjudgmental, peer-oriented manner. Most centers provided pamphlets and handouts to clients on various health topics and had libraries available for client use. The materials at feminist women's health centers differed from those provided by hospitals and HMOs in that they provided a wider range of options including mainstream medical information *plus* self-help and alternative treatments. Most centers also maintained an extensive list of referrals for mainstream, alternative, and psychosocial providers. Staffs were trained to provide information, referrals, and support for a woman's decision. A local newspaper article about the Richardson Women's Health Center focused on the center's philosophy of empowerment:

> The center and its workers are fiercely loyal to that philosophy—that a woman knows her body best and shouldn't take a doctor's command as gospel. That she be given plenty of time and attention, educated and allowed to make health decisions for herself. Even if that means an alternative to the traditional treatment. (TerHorst 1994)

Information and knowledge received at the feminist women's health centers were different from that obtained in mainstream settings in another way. Through demystifying medical procedures which have traditionally been done with women as passive recipients, women not only gained knowledge but were also empowered. In the feminist women's health centers, women were told step by step what was being done and why during an exam or procedure. Women were often encouraged to participate in their exams, for example, learning how to insert a speculum to see their cervix (what one respondent called "mirror and flashlight gynecology"), weighing themselves, and taking their own temperature and blood pressure. These are examples of ways in which knowledge gives patients more control and autonomy.

Most of the centers used lay health workers to provide a variety of services such as birth control counseling, blood pressure checks, prenatal care, and assisting in abortion procedures. The use of lay health workers helped reinforce the belief that women can learn how to take care of themselves from each other, thus increasing their confidence and sense of personal power. Even in centers where the use of lay health workers had decreased (typically due to regulatory and licensing issues), the philosophy of self-help and the support for women to be in control was still in place. In today's medical environment, knowledge does not always lead to power but it does lead to a more informed and, potentially, more assertive consumer.

Breaking Down Institutional Barriers

Institutionalized barriers prevent people from accessing the health care system or utilizing it effectively. Such systematic barriers include locating and getting to care, the impersonal and alienating atmosphere of many treatment settings, limited appointment time, the massiveness of bureaucracies and medical facilities, and the attitude of providers that efficiency is more important than the patient.

Feminist women's health centers attempted to break down institutionalized barriers and personalize service in several ways. They located themselves strategically near bus stops or locations easily accessible by public transportation. They chose offices in older, renovated homes or small office buildings rather than large, imposing medical complexes. Staff and physicians were often referred to by their first names and rarely wore lab coats or other uniforms, thus reducing the social distance between themselves and clients. As one client at Springfield Women's Health Center noted, "The Center itself provides a setting which is emotionally comforting—far from being cold and clinical." Another client remarked on the more personal atmosphere at the feminist health center, "Each counselor took the time to really look at the individual patients, to make eye contact. Too many places you go to today you get brushed off. You're just a number, a routine" (brochure from Springfield Feminist Women's Health Center).

One outcome of the current trends toward managed care has been a decrease in time clients spend with providers (Stichler, Phillips, McDaniels, Taubman 1996; Walker 1996), yet longer consultation time is essential to understand the entire social, psychological, and physical context of women's lives. The average length of appointments for an office visit at feminist women's health centers ranged from 15 to 60 minutes. The overall average was 22 minutes—more than twice as long as the 10-minute averages of many health maintenance organizations today (Anders 1996) and almost 25% longer than the 18 minutes found to be the minimum needed for participatory models of care (Kaplan et al. 1996, p. 502).

Feminist women's health centers also attempted to break down institutional barriers by providing special clinic nights for lesbians, having specially designed exam rooms for women with special needs, providing bilingual staff and resources, holding evening and weekend hours, and offering sliding fee scales. Through specialized services and attention to their clients' needs, the feminist women's health centers made care more accessible, more responsive, and more welcoming.

Another way to decrease institutionalized barriers to care is to take care out of the "institution" and situate it back into the community as neighborhood health centers did in the 1960s and 1970s. Besides locating facilities in the community (not within or next to a hospital), feminist health centers also provided community outreach programs. The extent of these programs varied but was considered an important aspect of their community and political work. During a visit to the Women's Health Collective, one long-time member talked about the presentation she was preparing for a Girl Scout troop:

> I'm going to give them little brown paper bags that have a speculum in them, directions on how to use the plastic speculum...an STD handbook...Probably a birth control pamphlet...and this little leaflet on self-treatment for yeast infections, and a chart on how you diagram symptoms of PMS. I've organized it so I'm meeting with the Girl Scouts specifically during one meeting and with the mothers in the second meeting. It's actually to give the mothers the same talk and same information but hopefully, the girls will feel a lot freer to ask questions.

Feminist women's health centers viewed community workshops as an important way to reach community members who might not normally access the health care system or who might be marginalized by the mainstream system. Information is presented in a safe, comfortable, "nonmedical" environment, and attendees are given tools, both physical and psychological, to take care of themselves. Providing preventive care in the client's own environment helps shift the locus of control toward the client.

The community often called centers for speaking engagements, for interviews with the press, or to be the spokeswomen for women's health issues. Centers generally provided these services free to community organizations and used them to help educate women about health issues and about alternatives to mainstream medicine. Some centers also ran their own educational programs or support groups, which were open to the public. Unfortunately, these programs were often the first to be cut when budgets were constrained.

Dignity and Respect

Another important aspect of the empowerment process was treating all women with dignity and respect. In order for clients to move toward empowerment, providers must treat them as though they can make choices that are right for them. Many health care consumers, particularly women, must first realize that they *deserve* to have their needs met in a health care interaction. Empowerment can only occur when women are ready to accept themselves as powerful and in control. The director at Colridge Women's Center, a center in a state with 25% of its residents on Medicaid, noted the emphasis they place on respecting all their clients:

> We are in a state with a lot of poverty but we will serve anyone regardless of income. There is an increased desire of all women to have a center that treats them respectfully. We treat them here, with women providers, and they get more time, particularly in our abortion service.

A thirteen-year member of the Women's Health Collective commented on the importance of a respectful environment at their center:

> The health center has functioned at least for its staff and I think a lot for its GYN clientele...as a safe place for women. It's a place where female process, however you define that or however it translates itself out, is preeminent. It's the given and it is a safe place to come to. It's a place where you can come in and be crazy about the world outside.

Dignity and respect flow easily from models of empowerment and patient-centered care. When providers focus on understanding the complete constellation of women's health, respecting the biological, psychological, and sociological factors, they dignify the clients' health or illness claims and empower them to be actively engaged in understanding and improving their health status.

EMPOWERMENT, QUALITY, AND MAINSTREAM MEDICINE

Feminist women's health centers prided themselves on the client-centered services they provided to women. They consciously built respect for each woman's situation and her decisions into their protocols and procedures. These centers operated on the assumption that women wanted knowledge and information and to be active participants in their health care decisions. Staffs were trained to be nonjudgmental, to help clients make decisions, and support them in those decisions. A member of a large West Coast center emphasized that "women walk out of here with more power." The importance of empowering clients was ingrained in feminist ideology and was passed on to clients in each encounter as staff gave them information, responsibility, and support for making their own choices.

Feminist women's health centers also consciously attempted to remove many obstacles to more open communications in medical interactions such as the setting, length of the medical encounter, and communication styles of providers. Providing a noninstitutional setting, allowing time for women to talk to providers who encourage dialogue, and decreasing the social distance between providers and patients are aspects of feminist models of care that have not been fully understood by many mainstream "women's health centers" who have co-opted the language of "empowerment" but have not truly given patients control and autonomy.

Several medical studies have now confirmed what feminist women's health centers have known for decades: giving patients/clients choices about, control over, and responsibility for certain aspects of care has important implications for patient loyalty, satisfaction with care, follow through with treatment decisions, and optimal health outcomes (Kaplan et al. 1996, pp. 502-503). These are characteristics of "quality" care that we must begin to assess. In order for mainstream medicine to recognize the benefits of feminist models of care, providers and insurers must begin to redefine "quality" and "cost-effectiveness." If cost-effectiveness continues to be defined by numbers of patients seen (market forces) rather than health outcomes and patient satisfaction, then we will continue to move away from a model of care that empowers patients and providers. As Kaplan and colleagues (1996, p. 503) concluded, "No amount of technically excellent care will produce optimal outcomes if patients are not actively engaged in managing disease, particularly chronic disease." Hart (1992, pp. 773-774) also reached similar conclusions, "We now recognize that patients supply at least 85% of the information required for diagnosis, and that their participation and understanding are essential for management of illness." Empowerment models are one way to foster engagement in one's own health care by encouraging patients/clients to "become experts in their own condition" (Hak and Campion 1997, p. 8).

The current emphasis on primary care offers the possibility for the development of more participatory models. Many aspects of feminist health models could be incorporated, particularly in community health settings, and can be cost-effective. For example, establishing clinics in churches, renovated homes, or nonmedical office buildings removes some barriers to access. Providing colorful patient gowns and mittens on stirrups and allowing patients/clients to be dressed when first meeting providers helps restore respect and dignity in the exam setting. Increasing the use of lay health workers helps decrease status inequalities and frees up other providers for more specialized work. Giving patients/clients more information about their exams and health status gives them more control over their health and respects their own knowledge. These strategies are all geared toward creating a more humane and respectful environment thus encouraging more participatory interaction between provider and patient. The key stumbling blocks will be constructed norms and beliefs regarding the nature of medical interactions and economic definitions of "cost-effectiveness." Empowering patients begins the process of redefining consumer expectations.

Central beliefs that need to be reconstructed begin with the disease model of health care. As long as patients and physicians believe that health is defined only as the absence of disease, the provider–patient interaction will be limited to conditions of illness, physicians will continue to be technicians repairing broken parts, and patients will remain passive recipients of the experts' instructions and recommendations. A window of opportunity has appeared as managed care and HMOs have placed primary care physicians into the role of gatekeepers and have expanded the use of allied health professionals. Having recognized the economic advantages of primary care and prevention over specialization and tertiary care, the next step is to begin to look at patient–provider interactions and "quality" of care. While most health care organizations have begun utilizing the rhetoric of prevention, education, and partnership with providers, few have actually implemented such programs. When consumers are treated with respect and dignity and begin to develop trusting relationships with providers, they will become more involved in their own care. Then mainstream health care organizations may begin to experience the same low rates of litigation that feminist women's health centers have.

Empowerment of patients/clients flows naturally from empowerment of workers. The benefits of empowering staff, in a patient-centered model, include increased job satisfaction, lower turnover, a greater sense of ownership in the program or facility, and improved continuity of care (Lathrop 1993). Feminist women's health centers created organizations that utilized and encouraged staff's talents and abilities and were flexible enough to respond to internal and external environmental forces. Some strategies utilized by feminist women's health centers to distribute power to workers included the following: (1) physicians utilized on a contractual basis, which made them "employees" accountable to the centers; (2) incorporating lay health workers, nurses, nurse practitioners, physicians assistants, and other nonphysician staff to minimize status differences between "providers" and "patients;" (3) discrepancies in the top and bottom levels of salaries were minimal and based mainly on seniority; and (4) workers were cross-trained to do more than one job to relieve boredom and provide back-up for one another. Structural components such as these were in place for ideological reasons (empowerment) but had the added benefit of helping to contain costs. Spending more time with clients is not necessarily more expensive (Hak and Campion 1997; Lathrop 1993).

Feminist women's health centers have continued to occupy a position just outside the mainstream medical system. This has given them the ability to design health care centers that are responsive to their communities and their clients rather than a corporation or stockholders. The institution of medicine is becoming increasingly "corporatized" focusing on efficiency and profit. As large corporations control more and more of the industry, communities lose control, patient's lose control, and ultimately, physicians lose control to the corporate executive office (Silver 1997). Critics claim that the mainstream model of health care is alienating, fragmented, and acute illness–oriented and that innovation can only occur within traditional health care delivery systems when providers begin to take into account individuals and the context of their lives, community-based issues, and the importance of prevention (McKenzie 1994). The surviving feminist women's health centers have a 20-year history of developing just such a health care delivery system and empowering patients/clients is at the heart of it. Many patients and providers are already attempting to redefine "cost-effectiveness" in terms of health, not simply profit, but the institution is slower to change.

More research is needed on participatory and patient-centered models of care, the link between these models and health outcomes, and how such models can be incorporated into primary care training programs and practice settings. However, it is also time to look outside the institution to community-based programs, which have been providing care without the constraints of mainstream medicine. Feminist models of empowerment in health care can contribute to the development of these new models, which enhance "quality" of care from the perspective of the patients and providers and not simply the perspective of the corporation.

ACKNOWLEDGMENT

A special thanks to Beth Rushing and Mary Zimmerman for their thoughtful comments and support and to the women of the feminist health centers for sharing their stories with me. An earlier version of this paper was presented at the American Sociological Association meetings, Toronto, Ontario, August 1997.

REFERENCES

Anders, G. 1996. *Health Against Wealth*. NY: Houghton Mifflin.

Bookman, A. and S. Morgen (Eds.). 1988. *Women and the Politics of Empowerment*. Philadelphia: Temple.

Foster, P. 1991. " Well Women Clinics—A Serious Challenge to Mainstream Health Care?" Pp. 79-94 in *Women's Issues in Social Policy*, edited by Mavis Maclean and Dulcie Groves. London: Routledge.

Gerteis, M., S. Edgman-Levitan, J. Daley, and T.L. Delbanco (Eds.). 1993. *Through the Patient's Eyes: Understanding and Promoting Patient-Centered Care*. San Francisco: Jossey-Bass.

Guttmacher, S. and J. Leeds. 1994. "Empowerment: A Term in Need of a Politics." Pp. 412-415 in *Beyond Crisis: Confronting Health Care in the United States*, edited by Nancy F. McKenzie. New York: Meridian.

Hak, T. and P. Campion. 1997. "Achieving a Patient-Centered Consultation by Giving Feedback in the Early Phases of the Consultation." Presented at the annual meeting of the American Sociological Association. August 10, Toronto, Ontario, Canada.

Hart, J.T. 1992. "Two Paths for Medical Practice." *The Lancet* 340: 772-775.

Kaplan, S., S. Greenfield, B. Gandek, W. Rogers, and J. Ware, Jr. 1996. "Characteristics of Physicians with Participatory Decision-Making Styles." *Annals of Internal Medicine* 124:497-504.

Lathrop, J.P. 1993. *Restructuring Health Care: The Patient Focused Paradigm*. San Francisco: Jossey-Bass.

McKenzie, N. 1994. *Beyond Crisis: Confronting Health Care in the United States*. New York: Meridian.

Merzel, C. 1994. "Rethinking Empowerment." Pp. 490-411 in *Beyond Crisis: Confronting Health Care in the United States*, edited by Nancy McKenzie. New York: Meridian.

Morgen, S. and A. Julier. 1991. "Women's Health Movement Organizations: Two Decades of Struggle and Change." Unpublished report.

Parsons, T. 1951. *The Social System*. New York: Free Press.

Pearson, C. and B. Seaman. 1998. "When Were We Founded? And Just What Were We Doing Back Then Anyway?" *National Women's Health Network News*, July/August.

Rodwin, M. 1994. "Patient Accountability and Quality of Care: Lessons From Medical Consumerism and the Patient's Rights, Women's Health and Disability Rights Movements." *American Journal of Law & Medicine* XX:147-167.

Ruzek, S.B. 1978. *The Women's Health Movement*. New York: Praeger.

_____. 1986. "Feminist Visions of Health: An International Perspective." Pp. 184-207 in *What is Feminism: A Re-examination*, edited by Juliet Mitchell and Ann Oakley. New York: Pantheon Books.

Silver, G. 1997. " Editorial: The Road from Managed Care." *American Journal of Public Health* 87: 8-9.

Stichler, J., K. Phillips, K. McDaniels, and J. Taubman. 1996. "Managing Women's Health in a Managed Care Environment." Presented at the annual conference of the National Association of Women's Health Professionals, Nov. 12, Chicago, Illinois.

TerHorst, C. 1994. "Demystifying Medicine." *Daily Herald*, March 31, section 4, p. 1.

Thomas, J. 1995. "Organizational Change in Feminist Women's Health Centers." Ph.D. dissertation, Department of Sociology, University of Colorado, Boulder, CO.

_____. 1999. "Ideology and Organizational Change in Feminist Women's Health Centers." *Gender & Society* 13: 101-119.

Walker, P. 1996. "Government Cuts and Rise of Managed Care Force a Medical Center to Shift Gears." *The Chronicle of Higher Education*, December 13, A30.

Weitz, R. 1996. *The Sociology of Health and Illness*. New York: Wadsworth.

Woolhandler, S. and D. Himmelstein. 1996. "Annotation: Patients on the Auction Block." *American Journal of Public Health* 86:1699-1700.

Zimmerman, M. 1987. "The Women's Health Movement: A Critique of Medical Enterprise and the Position of Women." Pp. 442-472 in *Analyzing Gender: A Handbook of Social Science Research*, edited by Beth Hess and Myra Marx Ferree. Beverly Hills: Sage Publications.

THE IMPACT OF FEMINISM ON MAINSTREAM MEDICAL SOCIOLOGY:
AN ASSESSMENT

Shirley Harkess

ABSTRACT

By its very nature, the field of medical sociology has considerable potential for incorporating a consideration of gender in its research. After approximately a generation of the women's movement, women's studies, and the study of gender in sociology, now is an appropriate time to assess the impact of feminism on the mainstream of the field. Such an assessment is distinct from a feminist critique of medical sociology, with its implication that the field has many shortcomings, and also from a review of the growing sociological study of women's health (e.g., Auerbach and Figert 1995) and their role in providing as well as receiving care. Instead, my purpose is to examine representative research in mainstream medical sociology for evidence of the extent and nature of feminism's influence. Overall, I argue that by the 1990s mainstream medical sociology has been significantly affected by feminism, but that this effect is qualified in important ways.

Research in the Sociology of Health Care, Volume 17, pages 153-172.
Copyright © 2000 by JAI Press Inc.
All rights of reproduction in any form reserved.
ISBN: 0-7623-0644-0

STATEMENT OF THE PROBLEM

The sociologies of family and work have been transformed by feminism and the study of gender. For sociology of the family, feminism electrified what was often viewed as a backwater of the discipline, which was less prestigious, at least partially because it was woman's (only) place. In the sociology of work the major societal change of women's postwar increase in employment made some response unavoidable. Is it possible to say something similar about medical sociology?

Two social movements growing from the larger women's movement—abortion rights and women's health—concern women's bodies specifically (see Zimmerman 1987). To these we could add the battered women's movement, another focal point of women's organizing, especially if we accept the critique that society treats family violence as a medical, not a social, problem. The rise of two or three very visible social movements is partly responsible for the impact of feminism on medical sociology. Without them there likely would have been much less. In addition, the status of women in the health care professions, particularly that of physician, has received considerable attention, although in the area of sociology of work. Despite these mobilizational focuses, however, the mainstream of medical sociology has changed less than the sociologies of family or work because the main object of its analysis remains the institution of medicine. Like this institution, although much smaller, mainstream sociology of medicine is highly organized in ways I will describe shortly that to date have limited the transformative potential of feminist social movements.

SOURCES AND CRITICAL METHOD

The basis for my assessment is an analysis of the 24 articles that appear in 1993 in the four issues of the *Journal of Health and Social Behavior*, the specialty journal in mainstream medical sociology. To these 24, I add the two medical sociology articles from the general mainstream of the discipline, as represented by the total of 83 articles, which appear in the *Annual Review of Sociology* from 1989 to 1993 and the *American Sociological Review* and the *American Journal of Sociology* in 1993. The titles of these 26 articles are in the bibliography.

To assess the impact of feminism on mainstream medical sociology, I determine the extent and mode of each article's treatment of women, gender, or feminism (Table 1). The *extent* of treatment is whether any one of these topics is mentioned at all and, if so, whether the reference is (1) a criticism, (2) a disclaimer stating why any or all of these issues will not be considered in any detail, (3) a passing mention, (4) an example illustrating some other issue, (5) a subsection of the article, or (6) a topic completely integrated throughout the article. I categorize the *mode* of discussion as (1) descriptive, (2) conceptual, or (3) theoretical (Table 1). In relating these two dimensions, it is unlikely, in practical

Table 1. Extent and Mode of Treatment of Women, Gender, or Feminism in Mainstream Medical Sociology

	Mode			
Extent	*Descriptive*	*Conceptual*	*Theoretical*	*Total*
criticism	0	0	0	0
disclaimer	1	0	0	1
passing mention	8	0	0	8
example of other issue	6	0	0	6
subsection of article	1	0	0	1
integrated throughout	3	3	0	6
Subtotal	19	3	0	22
no mention	—	—	—	4
Total				26

terms, that a disclaimer, a passing mention, or an example will be conceptual or theoretical.

Before I proceed with my assessment, it is important to note that mainstream medical sociology resembles the object of its analysis, the institution of medicine, in emphasizing scientific research as well as a model for that research. By way of illustration, all but two of the 26 articles feature technical statistical analyses of quantitative data.

One of the two exceptions is Becker's (1993) address to the Medical Sociology Section on receiving its Award for Distinguished Service to Medical Sociology at the 1992 meetings of the American Sociological Association (ASA). "A Medical Sociologist Looks at Health Promotion" is also the only article *on* the field itself; it does not, however, mention women or gender. The other nonquantitative piece is also the main one of three directly critical of the medical establishment. Conrad's (1992) "Medicalization and Social Control" appears, however, not in the specialty journal but in the general mainstream *Annual Review of Sociology*. (The other critical pieces are Takeuchi, Bui, and Kim 1993, and Yedidia, Barr, and Berry 1993). Conrad does argue, briefly, that "...women's natural life processes (especially concerning reproduction)" are much more likely than men's to be treated as a medical rather than a social problem (i.e., medicalized) and that "...gender is an important factor in understanding medicalization" (1992, p. 222).

In total, almost all of my selection of recent mainstream research on medical sociology (22 of 26 articles) is quantitative and not directly critical of the medical establishment. Supporting an impression of research in the natural sciences is the high incidence of multiple authorship, with six articles having more than three authors.

These 22 articles follow a model. Beyond four on quantitative methodology itself (Donovan, Jessor, and Costa 1993; Freedman 1993; Laumann, Gagnon,

Michaels, Michael, and Schumm 1993; Messeri, Silverstein, and Litwak 1993),
the model study concerns the relationship of some segment of the population to
some disease or condition. "Job Loss and Alcohol Abuse: A Test Using Data
From the Epidemiologic Catchment Area Project" by Catalano, Dooley, Wilson,
and Hough (1993) is a clear example. In the remaining 18 articles, the emphasis
is on the individual and what they do or do not do to maintain good health. Among
the diseases or conditions of interest to mainstream medical sociologists at this
time, the one of greatest importance is aging, with seven of 18 articles, because, I
suspect, it is both a condition and a population segment. The other 11 articles are
apportioned among stress, smoking, drinking, AIDS, and general health. Rather
than being a particularly "medical" list, it appears to be more "social" in that the
diseases or conditions more closely involve social processes, including those of
societal definition, than some others. This "social" emphasis is perhaps as it
should be, for medical sociology; however, it leaves the medical establishment
relatively more unscathed than would sociological analysis of an inherited disease
over which the individual has no control, as in the case of sickle cell anemia (see
Hill 1994).

INTEGRATING GENDER IN MAINSTREAM MEDICAL SOCIOLOGY

So, in the face of this structure, what enables me to claim that feminism has significantly affected mainstream medical sociology? The principal proof is that in
terms of the *extent* of treatment, only four articles, of the 26 total, do *not* mention
women, gender, or feminism in some way (Becker 1993; Simmons, Schimmel,
and Butterworth 1993; Cherry 1993; Turner, Hays, and Coates 1993) whereas six
integrate gender throughout the article (Table 1). In addition to such integration,
one article includes a subsection on gender; six use gender as an example; eight
mention women or gender in passing, and one enters a disclaimer. In terms of the
mode of discussion, all but three of these 22 articles at least mentioning gender are
descriptive (see Table 1). These three are conceptual. None of the 22 articles
engages in feminist theorizing. If this were the only way in which feminism was
defined, then you would have to say it receives no consideration here. In the
absence of feminist theorizing, I focus my analysis on the three integrative and
conceptual articles as the core of mainstream medical sociology's encounter with
feminism. I then pay some additional attention to the other integrative work and
lastly to the remainder of the articles. About one-third of the authors of all the
mainstream articles are women (26/73). With specific regard to the theme of this
volume, changing patterns of care and care provision—only one article in the segment of mainstream medical sociology I have examined considers it (Aneshensel,
Pearlin, and Schuler 1993).

CONCEPTUAL INTEGRATION

Three *Journal of Health and Social Behavior* articles (Robbins and Martin 1993; Conger, Lorenz, Elder, Jr., Simons, and Ge 1993; Fuller, Edwards, Sermsri, and Vorakitphokatorn 1993) constitute the core of mainstream medical sociology in 1993 conceptually integrating gender. Beyond this commonality, the articles are not on the same topic but are related in that the outcomes studied in the first two, drinking and stress respectively, are included among influences on women and men's health in the third. In contrast to all of the mainstream research in medical sociology, none of the work that conceptually integrates gender concerns aging.

In this core, Robbins and Martin's research on drinking offers the broadest view. Like some other researchers, the authors develop a gendered perspective in order to further understanding of an issue. In their case, it is sex differences (their term) in drinking behaviors and the concomitant fear that has been expressed: if women begin to get drunk as frequently as men and behave as they do when they drink, they will have more problems than men. The focus is on the presumed greater physical vulnerability of women to alcohol and the more negative reaction of others to intoxicated women than to men.

From their quantitative analysis of a nationally representative sample of 8,000 people, Robbins and Martin learn what women do to prevent those two specific problems: they do not become intoxicated as often as men, and they do not lose control as often when drinking. At least partly because they do not lose control, women are also no more likely to be criticized by others when intoxicated than are men. So it is *style* of drinking (control or not) more than properties of the substance or of the person that contributes to people's alcohol problems, an important point for medical sociology.

It is Robbins and Martin's contribution to introduce style of drinking, or control, from historical and crosscultural accounts of drinking behavior. They also operationalize the concept. Robbins and Martin introduce the concept of control as a possible explanation for sex differences in drinking based on what they describe as a "gendered deviance perspective." This perspective asserts that differences in sex roles (again, the authors' term) produce different forms of deviant behavior for women and men, and that, following Chodorow (1978), sex roles emerge from the socialization process. The gendered forms of deviance are internalization of distress for women and antisocial expression for men. Thus, for Robbins and Martin, socialization becomes the ultimate explanation for the finding that women exercise greater control when drinking. "Sex roles" (and socialization) also explain why others are no more critical of even women's *loss* of control than men's: others do not expect women to lose control, so they do not recognize it when it does occur.

Robbins and Martin do suggest, in their conclusion, that future research explore other factors or explanations:

Further gender role analyses could expand on the simple sex differences described here by exploring how these processes are affected by variations in sex role identification...and by specific sex role constellations of marital, family, and employment circumstances (1993, p. 317).

Such "constellations" would seem especially relevant to understanding why women do not lose control when they drink; that is, in terms of the authors' operationalization, why do women less frequently engage in aggression, become combative, drink to get drunk, or blackout when drinking than men? The question is what *are* the circumstances of women's lives currently (as adults rather than as children being socialized) that prevent or inhibit the abandon that men can afford or enjoy? Direct responsibility for the daily running of a household, including the care of dependents, would seem to be a major one, as Robbins and Martin surmise.

The other two articles in the conceptual integrative core of gender research in mainstream medical sociology address this issue—with varying, even surprising results. Unfortunately for purposes of precise comparison, the medical conditions they study differ: from Robbins' and Martin's drinking to stress (Conger et al. 1993) to general health (in Thailand) (Fuller et al. 1993).

Taking the opposite analytic tack from that displayed by Robbins and Martin in their work published in the *Journal of Health and Social Behavior* in 1993, Conger and colleagues (1993) *do* assemble a particular "constellation" of family statuses as the fullest expression of gender, at least in the U.S. They then examine processes at work *within* that construct. The authors sample 451 rural Midwestern white couples having all of the following characteristics: a primary breadwinner husband, a wife working outside the home no more than part-time, a seventh-grade biological child, and a sibling within four years of that child's age. In addition, the couples had been married an average of 18 years. To me, the characteristics of the sample suggest embodiment of the traditional gender ideal; that is, the way "it's supposed to be," or at least supposed to have been in the past.

With this group, Conger and colleagues examine depression, anxiety, hostility, and somatization as psychological distress in response to problems with jobs, money, health, the law, and family or friend relationships. Recognizing, importantly, that women and men report similar levels of distress overall, this study takes the hostility characteristic of men and adds it to, and distinguishes it from, past research on the anxiety and depression characteristic of women.

Two types of findings emerge: the expected presence of some gender differences and the unexpected absence of others (and thus similarity between women and men). Conger and colleagues explain the finding of expected gender differences in terms of socialization, as did Robbins and Martin previously, but oddly mislabel this on occasion as a social structural perspective (1993, p. 74). The unexpected absence of differences they attribute to the "very married" nature of the sample:

Given their life histories and location in family-oriented social systems, the husbands in this sample may well be as sensitive as their wives to negative family events... (Conger et al. 1993, p. 84).

Thus, ultimately, this "more traditional, relatively conservative family context" (1993, p. 85) explains both the differences as well as similarities among the men and women respondents. Although Conger and colleagues select their sample for the purpose of establishing a baseline of "traditional" gender *differences* in perceptions of specific stressors and responses to them, we learn that, in this repository of gender tradition (i.e., stable rural white Midwestern larger-than-average families), *similarities* develop between husbands and wives, a not unexpected result when family dynamics over the longterm are considered. In fact, such similarity may be an indicator of "happy families."

These findings lead Conger and colleagues to conclude that future researchers should vary the characteristics of their respondents as well as introduce measures of role responsibilities in the analysis. However, warning of a "shotgun" approach, they defend the strategy they chose:

Conducting such intensive research may require an emphasis on relatively homogeneous samples that can be subdivided by variables of theoretical interest (e.g., role identities) while being held constant on other significant factors (e.g., marital status). Such a strategy [i.e., theirs], compared to large-scale surveys of broad cross-sections of the adult population [e.g., Robbins and Martin above] may provide a cost-effective means for identifying the values, beliefs, and social processes that may help to account for gender differences in psychological distress (1993, p. 85).

In an analytic strategy that falls between Robbins' and Martin's omission of marital status and Conger and colleagues' complete elaboration of it, Fuller and colleagues (1993) sample married couples with one child and allow wives' employment and the number of preschool children to vary in the analysis. In addition, their study of gender and health in Thailand implements Conger and colleagues' suggestion of including (marital) role obligations. Like Robbins and Martin, the stated object of their research is to test the applicability of a "sex roles" explanation, in this case of morbidity differences between women and men in Thailand. Gendered morbidity is of such interest because it contradicts mortality patterns; that is, why/how are women "sicker" if men die younger? U.S. research implicates two acquired, as opposed to biological, risks—role obligations and psychological distress—as the main causes of gender differences in physical health (Fuller et al. 1993, citing Gove and Hughes 1979).

Fuller and colleagues operationalize marital role obligations as an irreducible minimum based on Gove and Hughes (1979):

When you are *really* sick, are you almost always able to get a good rest?...Even when you are really sick, are there a number of chores that you just *have* to do?... When you are *really* sick, is there someone to help take care of you? (1993, p. 258).

With a random sample of 2000 couples from Bangkok, they add as influences on health several variables resembling ones studied as outcomes in the other two studies: drinking problems, psychological distress, happiness, and emotional ties with spouse, neighbors, relatives, and friends. Despite these additions, gender differences in health are not muted, as we might expect from Conger and colleagues' results, but remain virtually unchanged, at least in this Bangkok setting. Thus, we have been unable to learn what it is specifically about gender that produces differences in health.

However, Fuller and colleagues, with no reference to previous research, introduce menstrual problems as another explanatory variable (1993, p. 258) measured as follows:

> During the past three months, have you had any of these symptoms: irregular periods? bleeding between periods? heavy bleeding during your period? unusually painful periods? being unusually tense or jumpy just before or during your period? (1993, p. 268).

With menstrual problems added to the analysis, it and psychological distress then emerge as significant influences on these Bangkok parents' health while, at the same time, the statistical effect of gender disappears. This suggests to Fuller and colleagues that menstrual problems are the main cause of Thai wives' poorer health and that experiencing psychological distress heightens women's perception of menstrual problems because the two are correlated.

For feminist and other perspectives in sociology, this is a precarious position. The authors argue that it is neither a role element of gender nor gender per se but menstruation that explains differential health in Thailand. Identifying menstruation as a (even *the*) "problem" for women, especially when it is also associated with psychological distress, is problematic because it recalls a biological explanation based on a distinguishing characteristic of women that has been viewed very negatively. In other words, it may confirm some people's "worst fears" or what others "knew all along"—that "that time of the month" made women sick if not also "crazy." Also, in the absence of attention to an analogous function for men, the study is an illustration of the greater medicalization of women's "natural life processes" that Conrad (1992) identified (see above).[1]

However, Fuller and colleagues' attempt to reinterpret this traditional view sociologically:

> Although menstruation has an obvious physiological basis, it also occurs within a cultural context. As such, the meaning attached to menstruation, the restrictions placed on menstruating women, and their responses to menstruation vary cross-culturally...This is not to say that menstruation is defined as illness in Thailand. Rather, these folk beliefs suggest that unless caution is taken, the symptoms of menstruation may be compounded... (1993, p. 265-266).

The researchers also argue that experiencing psychological distress exacerbates menstrual problems, rather than the stereotypic reverse. The influence of feminism in this research lies, then, in its aim to conceptualize (and operationalize) gender and in its sociological interpretation of menstruation.

This sociological interpretation of menstruation notwithstanding, the way menstrual problems are measured is questionable. The above questions about menstruation were apparently asked of the entire sample, men as well as women. Regardless, as Fuller and colleagues state, "By definition, men, who never have menstrual problems, have a score of zero" on the five possible symptoms (1993, p. 258). Thus, all 619 husbands in the study were coded one way (0), while only the 1398 wives could vary. It is important to note that it is in this way that the effect of gender is transferred to the effect of menstruation. While the procedure is technically correct, it may represent the victory of technique over reason. It harks back to the 1976 Supreme Court decision in *General Electric Company v. Gilbert* to uphold the manufacturer's claim that it was not discriminating against women when it denied insurance benefits for pregancy (while paying for hair transplants, a predominantly male procedure) but simply classifying employees as pregnant and nonpregnant persons. (This decision was overturned by Congress two years later with the Pregnancy Discrimination Act amending Title VII of the Civil Rights Act of 1964.) The disappearance of the effect of gender (with Fuller and colleagues' introduction of menstrual problems as a variable), then, is qualified, at the very least.[2] The original question remains: what is it specifically about gender in Bangkok (that varies for both women and men) that affects marital partners' health differently?

As assessment, if the three articles together are the principal evidence that feminism has transformed the representative portion of mainstream medical sociology I have reviewed, what is the nature of that evidence? Using the terms of my analysis (see Table 1), the articles do integrate a conceptualization of gender. The aim of each is to elaborate some aspect of gender as a way to explain differences in health between women and men: control of drinking behavior, exposure and response to negative life events, and marital role obligations, respectively. While the focus of each may seem narrow, the results do advance our understanding and raise questions for future research. The three articles are a core in the representative portion of mainstream medical sociology, which I have examined (26 articles total). I cannot, however, say that this core is sufficient to transform the mainstream completely.

I conclude this for four reasons, beyond the small proportion of the field this core constitutes (three of 26 articles). Firstly, since there is no explicit feminist theorizing, this conceptual core represents feminism's furthest reach in mainstream medical sociology. Indeed, the motivation for this research is not theory, as I am defining it, but the identification, measurement, and manipulation of specific health-related variables. Related to this reason is the fact that none of the 11 authors of the three articles is recognized as a gender scholar. Secondly, the conceptualization that does develop is oriented to role rather than social structure. Such an emphasis implicates socialization and ultimately the individual and his or her attitudes as the cause of gendered behavior and, by extension, as the mechanism of its change as well. This stance conflicts with feminist sociology's dominant structural approach.

Thirdly, the role that is addressed is marriage. With a sample of couples, Fuller and colleagues (1993) specifically, if minimally, test marital roles, and Conger and colleagues (1993) construct a sample of couples in order to embody gender tradition in marriage. Robbins and Martin (1993), neither specifying nor controlling marriage as a variable, essentially assume its effect as a constitutive element of gender (as well as identity) when they introduce controlled drinking to explain the gendering of this particular deviance. That is, because almost all men and women marry at some point, marriage is relevant, if not integral, to everyone's identity.

None of these three studies compares married men and women with unmarried (or divorced, separated, widowed, or never married) men and women to learn whether, to what extent, and how it is marriage that produces gender differences. Fuller and colleagues and Conger and colleagues could be building on the large body of previous research that has demonstrated that, at least in terms of health in the United States, it is better to be married than not and that, within marriage, husbands generally benefit more than wives. However, they do not say this. It would be interesting to discover the effect of marital status on the Thai population's health in Fuller and colleagues' study and on exposure and response to negative events in Conger and colleagues' study. Also, while marriage is the preeminent social institution structuring, and structured by, gender, there is more to gender than marriage—especially with more people marrying later and some not at all. Women and men outside of marriage are gendered subjects as well.

Fourthly, one of the three core articles is ambivalent about the role of biology. Some engagement with biology is understandable in the study of "sex differences" in general and in medical sociology in particular. In both cases, however, the relationship is generally meant to be subordinate. "The sociology of health and medicine" as an alternate label for the field, for example, indicates this: social factors influence the physical. For Fuller and colleagues (1993), though, biology takes precedent when it is introduced in the form of menstrual problems as a causal influence on Bangkok women's and men's health to clarify the meaning of gender.

Therefore, the feminist transformation of mainstream medical sociology is relative rather than absolute. It is, however, a "work in progress" because, taken together, these three articles make an unintentional contribution to the study of gender. Their different research designs address the conceptualization and operationalization of gender and its statistical analysis. The issues are perennial ones in social science research: Who is the sample?

What are the controls? And what are the variables? A substantive concern arises at this point: What does it mean to control for all sources of variation in order to produce a result of "no difference," especially no gender difference? In the assertion "but for these features, women and men would be the same," is the emphasis on what distinguishes the sexes or on the absence of difference? The answer is important for what it implies about the stance or objective of the research. Identifying distinguishing features of gender (the causes) implies a belief that inequality exists and a search for its diminution, whereas focusing on sameness (the result) implies a belief that essential equality exists and no action is required. In other words, has the researcher explained the way gender works, or "explained away" its effect?

In this core of mainstream medical sociology research on gender specifically, marked gender differences appear when there are few controls and a general sample of the population (Robbins and Martin 1993), while there are fewer or no gender differences and thus greater gender similarity when there are many controls and a sample of couples (Conger et al. 1993; Fuller et al. 1993 [when the latter add menstrual problems]). While Fuller and colleagues' conceptualization and implementation is flawed, as I discuss, they do attempt to investigate what it is in particular about marriage and about gender that may produce differences in women's and men's health, in this case. In their research, it is not just marriage but marital role obligations; it is not just gender but menstrual problems. Most importantly, the different analytic strategies followed by these researchers constitute a commentary on ways to study gender. They raise the issue of gender as a variable of degree rather than a category of "male" or "female."

DESCRIPTIVE INTEGRATION

Beyond this conceptual core of feminism in recent mainstream medical sociology, there are three other integrative articles. They are descriptive, not conceptual. They are also not *about* gender, at least as indicated by their titles. Instead, their topic is aging: "Are Black Older Adults Health-Pessimistic?" (Ferraro 1993), "Social Support from Friends and Psychological Distress Among Elderly Persons: Moderator Effects of Age" (Matt and Dean 1993), and "Kin and Nursing Home Lengths of Stay: A Backward Recurrence Time Approach" (Freedman 1993). Despite their titles, what leads me to categorize each article as integrating gender is their incorporation of its consideration. I suspect that given aging as the topic, it is impossible to avoid such consideration since gender differences in mortality and income visibly conspire to create an elderly population disproportionately composed of poor or near-poor women living alone. The three articles include sex/gender from the outset:

It is also possible that gender could be just as important as race in shaping health assessments (Ferraro 1993, p. 203).

A second variable of interest is gender because of consistent findings that men and women differ in certain types of social support and in the prevalence of psychological distress (Matt and Dean 1993, p. 188).

A final point often overlooked in studies of institutionalization is that the process is likely to be very different for older men and women…Because both kin patterns and the institutionalization process differ for men and women, our analysis is presented separately by gender (Freedman 1993, p. 140).

Briefly, in the context of these three articles, Freedman's purpose is more methodological, as her title suggests. In comparison, Ferraro and Matt and Dean narrow the focus to *men* in the interpretation of their results. Thus, in these two cases, paralleling developments outside of medical sociology, the evolving sociological interest in gender leads to discoveries about men. For Ferraro, for example,

A special consideration in assessing health status among the Black population that was further identified in this research is the unique situation of Black men (1993, p. 209).

In fact, the possibility of black men's selective survival is one motive for his work. His point of departure is the "racial mortality cross-over," which occurs at later ages when white deaths come to exceed black, rather than the reverse. With only men as the referent, however, the term "racial mortality crossover" is sexist. The author finds that older black men are not as sick as one would expect based on their race and gender:

Thus, this research contributes to a growing body of literature that reflects the unusually robust health condition of older Black men, presumably because of their selective survival to reach older ages (Ferraro 1993, p. 209).

With older black men less sick than expected, older black women are the most ill as well as the most pessimistic about their health, among the four race–gender combinations. Ferraro speculates that these differences are due to older black women's concentration on caring for others rather than themselves (1993, p. 210).

In comparison to Ferraro, Matt and Dean emphasize the *disadvantaged* condition of men, race unspecified, who are very old (over 70): "…(O)ld-old men in particular are especially vulnerable to psychological distress when losing friend support…" (1993, p. 187). The motive for this research is to map the effect of age on established gender differences in support and distress. Although the authors find that age intensifies the gender relation, they, like Ferraro, mention the implication for women only in their conclusion: "At the same time, old-old women are increasingly in danger of losing friend support because of their higher levels of psychological distress when compared to men" (1993, p. 198). Primarily because friend support appears more amenable to direct intervention than does distress (1993, p. 189, 198), Matt and Dean are more interested in speculating about the loss of social support with age and thus the increased importance of what support remains as the cause of men's problems, rather than about the greater psychological distress of older age as the cause of women's problems (1993, p. 197).

I welcome the discoveries of Ferraro and Matt and Dean about older men, but I also wonder at the fate of older women at the hands of mainstream medical sociologists, whether they are burdened, older black women or distressed, very old women. Will the analysis of their condition seem as pressing or as susceptible to successful intervention?

THE (NONINTEGRATIVE) REMAINDER OF MAINSTREAM MEDICAL SOCIOLOGY

Beyond these integrative descriptions of gender differences or the conceptual core analyzed first lies the remainder of recent mainstream medical sociology—descriptive and not integrative (see Table 1). One technical piece features a gender *subsection* (Aseltine and Kessler 1993). Six articles with *descriptive examples* concern medicalization (Conrad 1992), adolescent health (Donovan et al. 1993), minority youth in the mental health system (Takeuchi, Bui, and Kim 1993), elders' sense of efficacy (Grembowski et al. 1993), caring for elderly with Alzheimer's disease (Aneshensel and colleagues 1993), and televised health promotion (Flay, McFall, Burton, Cook, and Warnecke 1993). Eight articles only *mentioning* gender examine teen smoking (Ennett and Bauman 1993), the effect of drinking on employment (Catalano et al. 1993), elders' health perceptions (Johnson and Wolinsky 1993), support groups (Messeri and colleagues 1993), HMOs (Wholey, Christianson, and Sanchez 1993), and AIDS (LeBlanc 1993; Laumann et al. 1993; Yedidia, Barr, and Berry 1993). The distinction between an example and a mention is the amount of discussion—whether "some" or "very little." Finally, as I have defined extent of treatment, Pavalko, Elder, and Clipp (1993) issue a classic *disclaimer* as to why they do not include women in their study.

Conrad's previously noted essay (1992) on medicalization aside, the other five articles featuring a gender example and seven of the eight mentioning gender (including three in which it is simply an indicator of a sample's representativeness) do so mainly to dispense with it. That is, since "(i)n epidemiologic studies…sex has become a classic, standard potential risk factor of interest" (Matt and Dean 1993, p. 189), researchers must introduce gender in order to identify its effect, if any, on the outcome or process being studied. Having thus eliminated a source of potential criticism (if not also showing that gender has no significant effect), they are then free to concentrate on other variables and effects. This technique is similar to that used when gender itself is the focus: whether and how the introduction of statistical controls affects the relationship between the independent and dependent variables ("causes" and "effects") of interest. Takeuchi, Bui, and Kim illustrate well: "Since gender, age, and poverty status may help to explain minority–White differences in referral patterns…, they were included as control variables in subsequent multivariate analyses" (1993, p. 157). Flay and colleagues "...include gender in our analyses...Finally, we expected some of these [six] demographic variables to predict motivation to quit smoking as well as to predict viewing" (1993, p. 324). Since gender is an example or is only mentioned in these articles, by definition, it is not the focus of the research. Hence, whether differences appear as they do in almost all of these articles or do not appear is discussed very little (e.g., Johnson and Wolinsky 1993). In at least one case, it certainly seems that it should be.

In Catalano and colleagues' (1993) study of the effect of losing a job on alcohol abuse, gender has the same or greater effect as the main independent variable of interest (job loss) or as previous abuse of alcohol. The authors' focus on job loss obscures this fact. The conclusion of Donovan and colleagues shows the several ways this can happen:

> In our interpretation of the results of the second-order factor analyses carried out in the gender and ethnic/racial subsamples at each school level, we have tended to concentrate on the generalized consistency of the finding that a single underlying factor accounts equally well for the correlations among the health-enhancing behaviors in all subsamples, and on the consistency of the finding that all of the behaviors load significantly on that underlying factor in these subsamples...The fact remains that substantial variability still exists in the magnitude of those loadings across the ten subsamples. Future research clearly should delve more fully into the factors [like gender] accounting for this variability (1993, p. 358).

Since it is differences that attract in research, not similarities, one would think that gender would capture even more attention than it has in mainstream medical sociology. In discovering variables that differentially affect an outcome, researchers can make some attribution of cause. This notwithstanding, gender is not seen as "interesting" by everyone, a finding interesting in itself from the perspective of my assessment of feminism's impact on medical sociology.

One article among the six using gender as an example (and the only one on caregiving, the theme of this volume), Aneshensel and colleagues' (1993) "Stress, Role Captivity, and the Cessation of Caregiving," is actually *about* gender although the authors do not describe it in that way. I distinguish, as do other feminists, between gender as simply a variable and as a larger concept. Conceptually, this research concerns itself with gender, and with women more than men. However, it is virtually unremarked that two-thirds of the sample of those caring for impaired elderly relatives with Alzheimer's disease are women. Role captivity, the "construct that is the analytic cornerstone of this paper," is explained *in the conclusion* in terms of "people becom[ing] caregivers by default: because they are women, [are] not employed outside of the home, happen to live close by, and so on" (1993, p. 67). Earlier, Aneshensel and colleagues define role captivity as the involuntary incumbency of the caregiver role and operationalize it as a sense of entrapment (e.g., wish you were free to lead a life of your own) without referring to women/gender. Evidently, the role of gender in elder caregiving is so unremarkable because it is seen as commonplace. Thus, at least in this segment of mainstream medical sociology, the authors, in effect, accept the status quo. In doing so, there is a cost to men as well. With elder caregiving gendered "female," the relatively few men providing such care are overlooked, being treated, in a reverse of the usual, as "honorary women" for the purposes of the analysis.

Even as a variable instead of a concept, the effect of gender is minimized rather than explored in this study. As in the other articles with descriptive examples or mentions, Aneshensel and colleagues introduce gender as a variable to "cover their bases" and show that it has minimal effect on the causal processes that interest them. In doing so, however, they collapse instead of elaborate the pattern of gender variation. The six possible combinations of gender of elderly Alzheimer's patient and gender of caregiver (whether spouse or child) are reduced first to four and then further to two. The four gendered possibilities for the patient–caregiver relationship are husband caring for wife, wife caring for husband, daughter caring for mother, and all other parent–child dyads (1993, p. 59). This step eliminates the dyads of daughter–father, son–mother, and son–father. While Aneshensel and colleagues' reasoning is defensible on statistical grounds (to have meaningful, roughly comparable categories for analysis), some gender-relevant information is lost. Even more is lost when the variable is further reduced to whether the caregiver is simply a spouse or child. Here the reasoning is less defensible: no "reliable [statistical] association...[for] type of relationship between patient and caregiver" and the "dichotomy [is] used to facilitate stepwise procedure" (1993, p. 62, 64); the work on which these methodological conclusions are based is not shown. So, in Aneshensel and colleagues' shift from elaboration to dichotomy, what begins as gender, or gendered relationship, ends up simply as relationship. We do not learn, for example, how long daughters care for impaired mothers versus fathers and how they feel about it. Such knowledge could be important because of the contradiction the study uncovers: institutionalization relieves caregivers but hastens death for patients (1993, p. 54).

Finally, as an illustration of the lack of an explicit interest in gender, a last article in mainstream medical sociology in 1993 is a throwback to the "bad old days." Despite the unsuspiciously general and arresting title "Worklives and Longevity: Insights from a Life Course Perspective," in the abstract of the article, Pavalko, Elder, and Clipp (1993) unabashedly describe a much narrower study—of the standard middle-class men (the authors do not mention race) and the "career mobility" they are more likely to have. A brief acknowledgment of this lack appears in the text: "Our focus here is only on the men in the sample, leaving the more complex and varied worklives of the Terman women for future investigation" (1993, p. 366), which they may not do. The full, and depressingly characteristic, apology is relegated to a footnote to that statement:

We do this [exclude women] for several reasons. First, while women in the Terman sample were more likely than other women of their generation to pursue managerial or professional careers, only a small number had uninterrupted careers compared to the men. Even more important was the vastly different contexts in which these women pursued their careers. Contrasted with the men whose legitimacy of work identity was largely taken for granted, these women struggled simply to be accepted and recognized as professionals, thus defining issues of career progression in fundamentally different terms (1993, p. 377).

In light of the many, many studies that have been done of men and their career trajectories, it would be much more intriguing and potentially useful to us today to study these women's careers.

The Terman sample is composed of women and men born between 1900 and 1920 with an IQ greater than 135. It is a testament to the power of the above described research model in mainstream medical sociology that the temptation of studying such women would have been resisted. Although it probably is not the conscious intent of the authors, their disclaimer is very close to saying that not only are even these brilliant women not like men, but also they are not as "good" as men. While some may feel that it is necessary to establish the baseline before analyzing the deviant case, a signal accomplishment of feminism has been to invert what is figure and what is ground in any pattern. It is some token recognition of the impact of feminism that a disclaimer is entered at all. In contrast, in Aneshensel and colleagues, as previously noted, when women are two-thirds of the subjects, there is little acknowledgment that their study of caregivers to elderly spouses or parents with Alzheimer's disease is about women and women's lives.

Linking these two studies ignoring women and gender concludes this assessment on a disquieting note. Even more so when it is remembered that one of the authors of each of these two articles, Glen Elder and Leonard Pearlin, respectively, are major figures in sociology, not just medical sociology, collaborating here with somewhat less well known women (e.g., Aneshensel and Pavalko). Clearly, gender in all its ramifications remains to be fully explored in mainstream medical sociology.

CONCLUSION

To summarize, by the 1990s much in mainstream medical sociology considers women and gender. While the bulk of the field does not *concern* women and gender and feminist theory appears not at all, 22 of 26 mainstream articles published in 1993 (and Conrad 1992) mention women and gender in some way (see Table 1). A significant proportion of these—a half-dozen—integrate a consideration of gender into their analysis. The three whose integration is conceptual (Robbins and Martin 1993; Conger et al. 1993; Fuller et al. 1993) represent the extent of feminism's transformative effect on mainstream medical sociology. Their conceptualization is of gender as a role (like "sex role") rather than as a structure, and together they interrogate how one might study gender. This I consider an accomplishment of feminism in sociology, technical though it is, because, if the goal of feminism remains altering the mainstream, this is one place to begin: exactly how do you introduce gender into a statistical analysis and, yes, how do you operationalize gender?—as male–female, marriage, or menstruation, or some other way?

Some feminists object to the privileging of technique or method over conceptualization or theory. Are measurement and manipulation of gender as a variable *the* place to begin? In mainstream medical sociology, given its existing technical sophistication, I am arguing that it is. This is what people do in the mainstream of the field, in every substantive area, not just gender. Since women or gender is mentioned in 85% of the articles I analyze, I have had occasion to examine all but 15% of all mainstream articles. Some, feminists and others, may feel that this is quantification in the absence of conceptualization, their idea being that theory must be developed first before it can be tested. The response of mainstream medical sociology, as represented here, is that sufficient theory exists. Like the object of its study, medicine, mainstream medical sociology selects a part of the whole for intense scrutiny. The problem for feminism as opposed to medical sociology, though, is that the feminist theory that does exist is not being used. However, if feminists want to engage medical sociology, the exchange of ideas has begun, although the field has not yet been as thoroughly transformed as others.

If this, the measurement and manipulation of gender as a variable, is the main impact of feminism on mainstream medical sociology, is there a downside to the fact that women and gender appear in the majority of the research? I suggest that it is the introduction of gender only to get it out of the way. This occurs mainly in descriptive articles that simply mention gender or use it to illustrate another issue. There is a sense of the obligatory (as well as perfunctory) here; that is, it is not possible to publish research in mainstream medical sociology in the 1990s without at least taking into account (if only to bracket) the effect of gender—the possible difference that being a woman or man might make. This is a change that feminism has produced. While the sense of obligation implied is discouraging, the payoff—inclusion—is worth the cost. In the haste to dismiss gender, though, is a cost to mainstream medical sociology; the differences or similarities that are thus manifested are not discussed in any detail. Other impediments to the advance of feminism in mainstream medical sociology include remnants of the obliviousness and exclusion characteristic of the past, as represented in the work of some leaders in the field.

A last observation in presenting the qualified transformative effect feminism has had on mainstream medical sociology asks the question, what is *not* here? In other words, beyond the treatment of gender as a variable and some conceptualization, what else is possible? Is it a problem that there is very little recent research in what I have defined as mainstream medical sociology on some of the more obvious issues related to women such as childbirth, abortion, battering, and caregiving, as well as the women's health movement itself and women's status in the various health professions? Conrad (1992) objects, as has the women's health movement, that many of women's experiences and conditions have been medicalized, that is, defined, treated, and thus controlled as a problem and specifically a medical rather than social problem. If that is the case, should medical sociologists avoid these topics and thus the charge of furthering the medicalization of women's lives? It would appear that they are, unintentionally or otherwise, at least in the mainstream, notwithstanding a slight increase since 1993 in research on other topics specific to women or gender appearing in such publications as *Journal of Health and Social Behavior*.

Alternatively, instead of risking contributing to the medicalization of women's lives, research in mainstream medical sociology could also more frequently and critically analyze the medicalization process itself and the role of gender in it. In addition, sociologists could investigate why medical research does not and, in this way at least, insure that medical sociology *does*, adequately reflect the facts that, in the words of Bernardine Healy, President Bush's appointee as Director of the National Institutes of Health, women

- are the biggest consumers of health care,
- take the most medication,
- have the most operations,
- experience the majority of chronic diseases,
- and, at the same time, are the major caregivers in the home (in Maatz 1996, p. 1).

Although such research may appear in other journals in medical sociology, including ones in other countries, it is unlikely to appear in the mainstream *Journal of Health and Social Behavior* in the United States mainly because what I have found to be the standard research model of this journal does not easily lend itself to the type of critique I have suggested.

ACKNOWLEDGMENTS

A University of Kansas sabbatical leave provided assistance in the preparation of this chapter. An earlier version was presented at the 1998 meetings of the American Sociological Association.

NOTES

1.   In such a context it would be of interest to see the relationship between gender, marital role obligations, psychological distress, ejaculation, and health. Is it possible to imagine the following counterpart measure asked of male respondents:

During the past three months, have you had any of these symptoms: premature ejaculation? inability to ejaculate? unusually painful ejaculation? being unusually tense or jumpy just before or during ejaculation?

That it is not, at least in this particular type of study, and that the suggestion seems facetious if not ridiculous is only testament to Conrad's (1992) observation.

2.   There may also be some problem with the outcome variable of interest—health—and attempts to measure it in the Bangkok context. As the authors state:

Because Thais typically are not knowledgeable about specific diagnoses of the illnesses they experience, our health measures are necessarily general in nature (1993, p. 259).

This problem may be exacerbated by the complexity of the research design. In comparison to the previous two studies, which focused on drinking and stress respectively, this study casts these and other variable as intervening between the gender of the respondent and his or her ultimate general health status.

REFERENCES

Aneshensel, C.S., L.I. Pearlin, and R.H. Schuler. 1993. "Stress, Role Captivity, and the Cessation of Caregiving." *Journal of Health and Social Behavior* 34: 54-70.

Aseltine, R.H., Jr., and R.C. Kessler. 1993. "Marital Disruption and Depression in a Community Sample." *Journal of Health and Social Behavior* 34: 237-251.

Auerbach, J.D. and A.E. Figert. 1995. "Women's Health Research: Public Policy and Sociology." *Journal of Health and Social Behavior* (Extra Issue ["Forty Years of Medical Sociology: The State of the Art and Directions for the Future"]). 36: 115-131.

Becker, M.H. 1993. "A Medical Sociologist Looks at Health Promotion." *Journal of Health and Social Behavior* 34: 1-6.

Catalano, R., D. Dooley, G. Wilson, and R. Hough. 1993. "Job Loss and Alcohol Abuse: A Test Using Data From the Epidemiologic Catchment Area Project." *Journal of Health and Social Behavior* 34: 215-225.

Cherry, R.L. 1993. "Community Presence and Nursing Home Quality of Care: The Ombudsman as a Complementary Role." *Journal of Health and Social Behavior* 34: 336-345.

Chodorow, N. 1978. *The Reproduction of Mothering: Psychoanalysis and the Sociology of Gender.* Berkeley: University of California Press.

Conger, R.D., F.O. Lorenz, G.H. Elder, Jr., R.L. Simons, and Xiaojia Ge. 1993. "Husband and Wife Differences in Response to Undesirable Life Events." *Journal of Health and Social Behavior* 34: 71-85.

Conrad, P. 1992. "Medicalization and Social Control." *Annual Review of Sociology* 18: 209-232.

Donovan, J.E., R. Jessor, and F.M. Costa. 1993. "Structure of Health—Enhancing Behavior in Adolescence: A Latent-Variable Approach." *Journal of Health and Social Behavior* 34: 346-362.

Ennett, S.T. and K.E. Bauman. 1993. "Peer Group Structure and Adolescent Cigarette Smoking: A Social Network Analysis." *Journal of Health and Social Behavior* 34: 226-236.

Ferraro, K.F. 1993. "Are Black Older Adults Health-Pessimistic?" *Journal of Health and Social Behavior* 34: 201-214.

Flay, B.R., S. McFall, D. Burton, T.D. Cook, and R.B. Warnecke. 1993. "Health Behavior Changes through Television: The Roles of De Facto and Motivated Selection Processes." *Journal of Health and Social Behavior* 34: 322-335.

Freedman, V.A. 1993. "Kin and Nursing Home Lengths of Stay: A Backward Recurrence Time Approach." *Journal of Health and Social Behavior* 34: 138-152.

Fuller, T.D., J.N. Edwards, S. Sermsri, and S. Vorakitphokatorn. 1993. "Gender and Health: Some Asian Evidence." *Journal of Health and Social Behavior* 34: 252-271.

Gove, W. and M.D. Hughes. 1979. "Possible Causes of the Apparent Sex Differences in Physical Health: An Empirical Investigation." *American Sociological Review* 44: 126-146.

Grembowski, D., D. Patrick, P. Diehr, M. Durham, S. Beresford, E. Kay, and J. Hecht. 1993. "Self-efficacy and Health Behavior among Older Adults." *Journal of Health and Social Behavior* 34: 89-104.

Hill, S.A. 1994. *Managing Sickle Cell Disease in Low-Income Families.* Philadelphia: Temple University Press.

Johnson, R.J. and F.D. Wolinsky. 1993. "The Structure of Health Status Among Older Adults: Disease, Disability, Functional Limitation, and Perceived Health." *Journal of Health and Social Behavior* 34: 105-121.

Laumann, E.O., J.H. Gagnon, S. Michaels, R.T. Michael, and L.P. Schumm. 1993. "Monitoring AIDS and Other Rare Population Events: A Network Approach." *Journal of Health and Social Behavior* 34: 7-22.

LeBlanc, A.J. 1993. "Examining HIV-related Knowledge among Adults in the U.S." *Journal of Health and Social Behavior* 34: 23-36.

Maatz, L.M. 1996. "Dean Discusses Women's Health." *Women's Studies Newsletter.* Summer: 1. Columbus, Ohio: The Department of Women's Studies, The Ohio State University.

Matt, G.E. and A. Dean. 1993. "Social Support from Friends and Psychological Distress among Elderly Persons: Moderator Effects of Age." *Journal of Health and Social Behavior* 34: 187-200.

Messeri, P., M. Silverstein, and E. Litwak. 1993. "Choosing Optimal Support Groups: A Review and Reformulation." *Journal of Health and Social Behavior* 34: 122-137.

Pavalko, E.K., G.H. Elder, Jr., and E.C. Clipp. 1993 "Worklives and Longevity: Insights from a Life Course Perspective." *Journal of Health and Social Behavior* 34: 363-380.

Simmons, R.G., M. Schimmel, and V. Butterworth. 1993. "The Self- Image of Unrelated Bone Marrow Donors." *Journal of Health and Social Behavior* 34: 285-301.

Robbins, C.A. and S.S. Martin. 1993. "Gender, Styles of Deviance, and Drinking Problems." *Journal of Health and Social Behavior* 34: 302-321.

Takeuchi, D.T., K-V.T. Bui, and L. Kim. 1993. "The Referral of Minority Adolescents to Community Mental Health Centers." *Journal of Health and Social Behavior* 34: 153-164.

Turner, H.A., R.B. Hays, and T.J. Coates. 1993. "Determinants of Social Support among Gay Men: The Context of AIDS." *Journal of Health and Social Behavior* 34: 37-53.

Wholey, D.R., J.B. Christianson, and S.M. Sanchez. 1993. "The Effect of Physician and Corporate Interests on the Formation of Health Maintenance Organizations." *American Journal of Sociology* 99: 164-200.

Yedidia, M.J., J.K. Barr, and C.A. Berry. 1993. "Physicians' Attitudes toward AIDS at Different Career Stages: A Comparison of Internists and Surgeons." *Journal of Health and Social Behavior* 34: 272-284.

Zimmerman, M.K. 1987. "The Women's Health Movement: A Critique of Medical Enterprise and the Position of Women." Pp. 442-472 in *Analyzing Gender: A Handbook of Social Science Research*, edited by B.B. Hess and M. Marx Ferree. Newbury Park, CA: Sage.

BIO–POWER AND RACIAL, CLASS, AND GENDER FORMATION IN BIOMEDICAL KNOWLEDGE PRODUCTION

Janet K. Shim

ABSTRACT

The inclusion of race/ethnicity, socioeconomic status, and sex/gender in biomedical and epidemiologic research often constitutes routine and taken-for-granted practices that are based on particular notions of bodily "differences" and their roles in health and illness. Such practices legitimate constructions of race, class, and gender as attributes of atomistic individuals—rather than as intersectional dimensions that structure social relationships—and render invisible how relations of power contribute to the stratification of well-being and disease. This paper offers applications of two theoretical perspectives to illuminate these arguments. Firstly, epidemiologic research exemplifies in many ways Foucauldian notions of bio–power and Panopticism. It individualizes bodies and bodily differences; at the same time, it disindividualizes power, embedding it within the diffuse and pervasive acts of biomedical knowledge production and subsequent imperatives of self-

Research in the Sociology of Health Care, Volume 17, pages 173-195.

judgment and surveillance. Secondly, epidemiologic research embodies processes of racial, class, and gender formation, and constitutes a kind of racial, class, and gender project. Such projects mediate between the discursive definitions of bodily differences and the institutional forms in which those definitions are routinized and standardized. As such, biomedical knowledge production is an active participant in the construction and institutionalization of social meanings of "difference." However, my contention is not that we should abandon the epidemiologic use of racial, class, and gender categories. Instead, race, class, and gender must be reconceptualized as social relations of power that are located not just in the biological bodies of individuals but in the social spaces between them, producing and stratifying the distribution of health and illness.

INTRODUCTION

Race, class, and gender have been focal points for biomedical[1] inquiry throughout U.S. history. Coinciding with increasing attention to medical issues and a growing awareness of the possibilities of biomedical science throughout the nineteenth and twentieth centuries, a scientific body of knowledge about the effects of racial, class, and gender differences on health has been evolving. Currently, the predominant mode in which measures of race, class, and gender are incorporated into biomedical research is as part of a medley of demographic information collected in conjunction with data on health status, interactions, and outcomes. For example, racial, sex, and socioeconomic categories are used to characterize a study population, and to adjust or stratify results of multivariate analyses (LaVeist 1996). Such biomedical research has revealed that these factors are associated with variations in the incidence and prevalence of disease, and with patterns of health care utilization and health beliefs.

The underlying assumption is that racial, class, and gender categories have some biomedical significance—that taking them into account in producing biomedical knowledge is scientifically meaningful. Certainly the material significance of racial, socioeconomic, and gender differences in health interactions and outcomes demands their inclusion in biomedical, epidemiologic, and public health research. However, the implications of such inclusion are rarely discussed explicitly in those literatures. Instead, many debates have focused on methodological issues, particularly the difficulties of measuring or operationalizing race and class (e.g., Cooper 1994; Hahn and Stroup 1994; McKenney and Bennett 1994; Schulman, Rubenstein, Chesley, and Eisenberg 1995), the causal mechanisms through which such characteristics impact health (e.g., Nickens 1995), and more recently, the conflation of sex and gender.[2]

Against these debates, I argue that the practices by which dimensions of race, class, and gender are conventionally taken into account in health research legitimate constructions of those dimensions that are individualistic and essentialist, and further their routine and institutionalized use. Accordingly, they reduce *socially* produced health inequalities to the manifest outcomes of *individual* differences in risk factors, and thereby obscure the role of *relations of power* in the construction of biomedical knowledge and stratification of well-being and ill health. Thus biomedical knowledge production practices that categorize individual bodies by race, class, and gender contribute to reifying and sustaining inequalities in health and illness along those dimensions, rather than illuminating potential remedies.

In this paper, I first examine current biomedical models of disease causation and epidemiologic practices as the dominant theoretical and methodological foundations for producing knowledge about the health effects of racial, class, and gender differences. I do so in order to focus the discussion on how such bodily "differences" are constructed and mobilized in biomedical explanations for illness distribution. I then contrast a constructionist understanding of "difference" against the interpretations that are dominant in current biomedical knowledge production. Two theoretical perspectives are used to illuminate the meanings of biomedical constructions of race, class, and gender. First, Foucauldian notions of bio–power and Panopticism (Foucault 1975, 1977, 1978, 1980) lend critical lenses through which to analyze epidemiologic practices and implications. Second, the concepts of Omi and Winant (1994) of racial formation and racial projects provide a powerful basis for understanding how such practices work and how they can come to have oppressive consequences. Finally, I conclude with some observations about the repercussions of dominant knowledge production for addressing health inequalities and social change.

BIOMEDICAL MODELS OF DISEASE CAUSATION AND PRACTICES OF EPIDEMIOLOGY

This paper is situated within a larger project that pivots around two fundamental questions: How is the distribution of health and illness among individuals and populations explained? What bodily features, characteristics, and differences are conceptualized to be relevant in such explanatory accounts? Based on these questions, I interrogate the kinds of knowledges and practices that have been used to construct and sustain an intelligible story about the presence of health and illness among human populations. Such explanatory accounts of the presence and distribution of disease implicitly or explicitly construct, as well as reflect, underlying conceptions of disease causation and etiology.

Currently, the dominant explanatory framework in the United States is that of biomedicine, within which the *multifactorial model of disease causation* constitutes a central and widely used paradigm. The multifactorial model posits that most illnesses are the result of multiple sources of causation, involving a complex interplay of agent, environmental, and host factors[3] (Evans 1978; MacMahon,

Pugh, and Ipsen 1960; Susser 1985). The incidence and distribution of disease is viewed not as random occurrences but as linked to specifiable factors of suscepti- bility and exposure. Thus, contemporary biomedicine understands the health sta- tus of individuals to be the outcome of their particular constellations of health risks and exposures.

The discipline primarily responsible for the identification and verification of these risk factors is that of epidemiology. Epidemiology is the scientific field con- cerned with disease patterns in human populations and with the factors that influ- ence these patterns; its fundamental project is to develop predictive models of health status. Epidemiology serves as the dominant framework for understanding risk in biomedicine, where risk refers to the estimated excess frequency of an occur- rence in a population (e.g., Ginzburg 1986), and is usually presented as a statement of statistical probability. These probabilities provide a quantitative basis for deter- mining correlation between two series of events (Ginzburg 1986)—for example, the correlation between race and excess frequency of, or risk for, a disease.

Epidemiologic data on these risks provide a wealth of knowledge used to legit- imate clinical intervention into the behaviors of patients with the aim of making the enterprise of predicting health outcomes and promoting health more rational. Despite the fact that the statistical identification of risk factors most often does not demonstrate causality nor predict individual outcomes (a fact often obscured in public reports of study results), epidemiologic findings do affect medical inter- vention and clinical decision making (Armstrong 1995; Becker and Nachtigall 1994; Williams 1997). As the practices of individual risk assessment and health promotion[4] demonstrate, a consideration of epidemiologic patterns is believed to constitute a more scientific and efficient basis for identifying individuals whose characteristics and/or behaviors subject them to higher risk for disease. Such pat- terns of risk also provide the evidence on which the significance of the role of racial, class, and gender differences in disease causation can be affirmed. Current risk constructions of race, class, and gender are therefore claimed to be the ratio- nal outcomes of the scientific practice of epidemiology and its application in clin- ical practices.

However, this surface view of knowledge production about race, class, gender, and risk, and its implementation in medical practice, is problematic and contest- able in many ways. Of special note here, most epidemiologic findings emerge from research in which the unit of analysis is the individual. As a result, epidemi- ology regularly examines agent, environmental, and host dynamics of disease causation only at the level of the individual and simultaneously simplifies a com- plex world into discrete, presumably independent units of observation. "Problems of scale, complexity, and multidimensionality have thus been minimized" (McMichael 1995, p. 633). The theoretical image of a complex and intercon- nected "web" of both causal and protective factors (MacMahon, Pugh, and Ipsen 1960) that together determine an individual's health status in practice focuses attention on those risk factors *closest* to the outcome of interest. These typically translate to the "direct" biological causes of disease and to the lifestyles or behaviors addressable at the individual level (Krieger 1994).

Some "social" factors—commonly referred to as psychosocial, behavioral, and demographic variables—are considered to contribute to health outcomes. However, they are also viewed as individualistic attributes and even their relevance for disease causation, particularly given the rise of genomics,[5] seems to be quite secondary to biological and physiological dynamics (Krieger 1994; McMichael 1995). Moreover, the fairly routine finding that many of these psychosocial and behavioral factors do indeed differ by race, socioeconomic status, and sex reinforces the notion that not only race, class, and gender, but also other contributors to health, can be validly conceptualized as discrete and static attributes of the individual host. Connected to this is the assumption that "environmental" factors, as incorporated in the multifactorial model, are exogenous to the individual; that is, one's circumstances are taken as a given, as if individuals were dropped into a set of conditions that are not socially constructed nor patterned (Krieger 1994).

Such consequences further *concretize* the notion that nonintersectional and individualistic constructions of racial, class, and gender differences help "explain" the distribution of health and illness, pushing aside uncertainty and ambiguity over what *exactly* about one's race, class, and gender contributes to disease. Thus modern epidemiology—despite nascent and spotty efforts to develop a more ecological model that acknowledges causation at multiple levels (Susser and Susser 1996)—remains predominantly concerned with the identification of individually conceptualized risk factors (Pearce 1996). By representing race, class, and gender as characteristics of individuals, biomedicine implicitly asserts that those attributes, and more importantly their impacts on health and illness, do not change despite shifts in definitions, social policies, cultural norms, and other social and historical contingencies. Given these reductionist and individualizing tendencies of biomedical research practices, I argue that their constructions of racial, class, and gender differences are *essentialist*. Following Omi and Winant (1994, p. 187), I define essentialism as "belief in real, true human essences, existing outside or impervious to social and historical contexts."

Thus, rather than being an objective and neutral observer of some "true," natural reality, biomedicine *constructs* knowledge about the health effects of racial, class, and gender differences; in the process, it necessarily engages in reflecting, sustaining, and shaping social meanings of "difference." The constructions and mobilizations of race, class, and gender in biomedical explanations for disease etiology and distribution in turn posit particular kinds of social relationships and obscure others and render some implications for health practices far more likely than others. Specifically, the individualistic and atomistic constructions of race, class, and gender typical of biomedical and epidemiologic research endeavors render invisible the social relations of power that contribute to the stratification of health and illness. I believe that to address such quandaries and to better understand the relevance of racial, class, and gender differences for disease etiology and distribution, we must reconceptualize those differences as the products of ongoing processes of social construction and interaction, and as expressive of socially constructed relations of power. In the following section, I propose some possible reconceptualizations.

CONSTRUCTIONIST APPROACHES TO "DIFFERENCE"

Using a social constructionist approach, I define race as a category of identity and social stratification that refers to phenotypic differences, conflated with assumptions about differences in cultural and geographic origin and the significances attached to each of these kinds of distinctions. The meanings, including biomedical ones, given to race as a general concept or to particular races have been reconstructed over time and across social contexts. These meanings, which shift in terms of their content and salience, have included definitions of superiority and inferiority, notions of class and socioeconomic status, ethnic and cultural differences, national ancestry, and biological and genetic traits. Constructions of race continue to be crafted, sustained, and mobilized to protect dominant group interests[6] and to manage potential conflicts; race therefore contributes to the organization of human bodies in social spaces and relationships. More specifically, to the extent that individual bodies categorized as *racially* distinct are assumed to be *essentially* different, racial inequities in health and illness can be explained away as the "natural" and "rightful" order of things.

Similarly, I understand gender as the organization of social relations among individuals and groups based on supposed differences of sex. This definition rejects a biological determinism, instead emphasizing the constructed yet consequential nature of gender differences that are defined relationally and in interactions. By focusing on the social processes and actions that reify the distinctions between the "male sex" and the "female sex," my understanding of gender contests both the classification scheme of two opposite and essential sexes[7] and the criteria by which individuals are categorized within it. Gender, in my view, is a historical and social construction whose classificatory schemes, criteria, and meanings are constantly policed and negotiated. As one authoritative arbiter of biological matters, biomedicine is therefore critically positioned to define gender categories as manifest in the biologies of bodies and to link those categories with health outcomes, thereby investing gender categories with meanings and consequences.

Finally, I argue that social class can best be understood as a relational attribute that refers to a group's position within economic structures and processes. Class is thus a social relationship of power that is marked by location within modes of economic production, distribution, and consumption. Class categories, like those of race and gender, are continually renegotiated and therefore always in some measure of flux. Important elements of class have included measures of income, wealth, occupation, social status, economic power, and education. However, social class, when invoked in biomedical knowledge production, is most often constructed simplistically as socioeconomic status (SES). Unlike sex, class seems to be understood as an analytic category having to do with largely social dynamics and, at times, serving as a proxy for cultural differences. Yet at the same time, biomedical constructions of class, while being thought of as "social," are not explicitly conceptualized as *relational*. Various aspects and measures of class are mobilized to impose categorical distinctions between classes; when these are conflated with behaviors and attributes perceived to be class-specific, what result are deterministic and essentialist notions that legitimate exclusionary and discriminatory practices. Moreover, the class stratification of multiple kinds of outcomes, including health status, is—as with race—interpreted as the logical and natural product of class-specific behaviors and traits and therefore viewed as unproblematic.

Importantly, definitions of and beliefs about race, class, and gender have always been mutually constituted in reference to each other. Race and class, for example, have particularly acute interactions and intersections. This can be evidenced by the incommensurability of class indicators among different races: the increments to income and wealth with each additional unit of education have been shown to be lower for African Americans than Whites (Oliver and Shapiro 1995; U.S. Bureau of the Census 1993); additional units of income in predominantly African American neighborhoods have lower purchasing power than in largely White communities (Villemez 1980; Williams and Collins 1995); and the health benefits of additional units of income are far greater for the poor than for the affluent (Guralnick and Leveille 1997). Yet race and class have been conceptually conflated in health research in ways that provide little insight into either their intersections or theoretical distinctions. Consequently, embedded in the routine use of biomedical constructions of race and class to account for variations in health status and service utilization is the notion that being non-White and/or poor are *individual–level* and *additive* risk factors for adverse health outcomes. Conversely, being White and of middle or upper socioeconomic status are generally and implicitly regarded as protective to health. Most critically, the "health risks" of being a person of color and/or poor are seen to be unrelated to the usually ameliorative effects of being White and middle-/upper-class. That is, race and class, and their subsequent effects on health, are seen to be individual rather than relational and social phenomena.

The intersections between gender and both race and class in health and illness are equally profound. For example, "hysteria" has always been a gendered medical phenomenon, but more specifically it was almost exclusively an affliction of the middle and upper classes of White women (e.g., Lupton 1994). While the construction of womanhood, aided by biomedicine, positioned wealthy (usually White) women as dependent and fragile and therefore barred them from public roles and physical activities, working class and poor women and women of color "enjoyed few of the dubious benefits of the ideology of womanhood" (Davis 1981, p. 5). Because the work of these groups of women was indispensable to the needs of a growing capitalist economy, they were seen as exempt from restrictions on heavy labor. In a more contemporary example, the regulation of reproduction—not just explicitly through policies regarding reproductive choice and autonomy but also through strategies ranging from family and welfare policies, health care delivery, to birthing practices—has also been acutely oppressive along racial and class (not to mention gender) lines (Bayne-Smith 1996; Davis-Floyd 1992; Martin 1987; Roberts 1997).

In short, I argue that beliefs around race, class, gender, and their categorizations are socially, politically, and culturally defined and performed in interactions. Furthermore, they are intersectional social processes: race, class, and gender mutually transform each other within lived experiences and constitute constructions of bodily differences that shape and reflect social relations of power. However, more often than not, measures of race, class, and gender are separated in epidemiologic analyses and constructed in ways that justify essentialist understandings of bodily "difference," obscuring the ways in which relations of power are foundational to their construction. All too often in biomedical research, race, class, and gender are implicitly conceptualized as *distinct* and *individual traits* rather than as intersectional dimensions that structure social relationships. That the health disadvantages of being of a particular race, class, or sex are not seen as linked to the advantageous effects of belonging to other groups highlights the existence of unspoken norms in biomedical knowledge production—White, male, middle- or upper-class—against which all other groups are compared. This construction of center/margin dichotomies implicitly acts to mark all other types of bodies, to reinforce the representation of associated behaviors—by virtue of their association—as "deviant," and to again obscure the social relations of power that are embedded within these dichotomies.

As biomedical research practices propose and legitimate essentialist constructions of race, class, and gender, they also defend an understanding of disease as an individual phenomenon while dismissing the notion that health and illness are products of social relationships. These practices and the knowledges they produce then help to structure biomedical interventions and health policies that are ostensibly aimed at ameliorating health inequities, but with limited efficacy and questionable utility. In short, interventions modeled upon individualistic constructions of race, class, and gender help to sustain dominant social processes and relations of power that in turn are embodied, producing inequalities in health status. By circumscribing the causes of health and illness to individual-level racial, socioeconomic, and sex differences, and ignoring the *constructed* nature and *relational* dynamics of these differences, biomedical knowledge production as currently practiced is limited in what it can do to further health equality and social justice.

I now turn to the first of two theoretical approaches that help to illuminate and extend these arguments: Foucault's notion of bio–power and his assertions about the intimate linkages between knowledge and power. This discussion is then followed by an exploration of Omi and Winant's (1994) racial formation theory, their concept of racial projects, and the implications their ideas have for considering biomedical constructions of race, class, and gender.

BIO–POWER AND THE INDIVIDUALIZATION
OF BODIES

Foucault's (1975, 1977, 1978, 1980) works are marked by his ongoing concern with the emergence of a new kind of "microphysics of power" in eighteenth- and nineteenth-century Western societies. During this period, he argues, dominant mechanisms of power shifted from coercion and ritualized public punishment to surveillance and normalization. He views this shift not as a curious historical accident, but as the direct result of the construction of various "problems of population" (e.g., birthrate, longevity, public health, and housing), the simultaneous development of several fields of knowledge, and the refinement of more diverse and efficient techniques of human control (Foucault 1978). Governments perceived that they were dealing not merely with individual subjects, but with a population characterized by specific phenomena and its own particular economic and political problems. This new concept of "population" constituted, in Foucault's mind, one of the innovations that helped to usher in the modern era of *bio–power*.

While the exercise of power aimed at disciplining the population was localized in various institutions (such as the school and the hospital), it was also the objective and the effect of new types, or disciplines, of knowledge such as demography, public health, statistics, and medicine. In these new fields, power and knowledge can be seen as mutually productive: "the formation of knowledge and the increase of power regularly reinforce one another in a circular process" (Foucault 1977, p. 224). Foucault captures the nature of this relationship in his concept of *power–knowledge*. Bio–power referred to a new form of power–knowledge as "an agent of transformation of human life" (Foucault 1978, pp. 142-143), engaged with the regulation and distribution of human bodies in the domain of value and utility.

In particular, the application and exercise of bio–power leads to the description, comparison, arrangement, and distribution of individuals. Bio–power fixes and renders individual differences "scientific," pinning individuals—arrayed around constructed notions of the norm—down in their particularity. Individuals become "cases," to be used for domination, constituting *both* the object and the effect of power–knowledge relations: "power and knowledge...invest human bodies and subjugate them by turning them into objects of knowledge" (Foucault 1977, pp. 27-28). Foucault accordingly rejects the thesis that this period of modernism and bourgeois capitalism gradually effaced the body and denied its biological and physical reality in favor of consciousness, ideality, and the soul (Foucault 1978, 1980). In fact, he argues, nothing was more material or corporeal than the development of modern disciplines and the exercise of technologies of power that take life and the body as their central target and object: "the historical moment of the disciplines was the moment when an art of the human body was born...The human body was entering a machinery of power that explores it, breaks it down and rearranges it... produc[ing] subjected and practised bodies, 'docile' bodies" (Foucault 1977, p. 138). Disciplinary technologies invest what is most material and vital—the body, its physiologies, and its behaviors—with meanings and values (Foucault 1978). Power, in contemporary society, is therefore seen as a "political anatomy of the body" in which apparatuses and technologies exert diffuse yet constant forces of surveillance and control on human bodies and their behaviors, sensations, physiological processes, and pleasures.

These arguments invite a consideration of the power–knowledge relations behind various discourses on health risks, and the discursive connections that are made to concepts of race, social class, and gender in biomedicine and epidemiology. The underlying logic of epidemiology is the comparison of categories of individuals characterized by particular demographic, behavioral, and other factors, and the determination of which factors are associated with excess morbidity and mortality. To the extent that race, socioeconomic status, and sex enter into the analysis (and they do so routinely in contemporary epidemiologic practice), individuals are grouped into categories and often homogeneity within categories is assumed. Measures such as race, socioeconomic status, and sex implicitly serve as proxies for, and presume the existence of, genetic, behavioral, and other "essential" determinants of susceptibility linked to racial, class, and gender differences. Thus the normative epidemiological model, as I argued earlier, is an additive and reductionist one in which race, socioeconomic status, sex, and other factors are decontextualized and viewed as characteristics of individuals.

The construction of epidemiologic knowledge and its application in clinical practice can thus be seen as an example of a power–knowledge relation: the production and centralization of knowledge in order to maximize control over individual and social bodies. By understanding and studying humans as atomistic and individualized bodies, biomedicine inevitably structures the research questions that are asked and the kinds of answers that are sought, sustaining a diffuse kind of bio–power. Specifically, inequalities in health status and the characteristics found to be associated with them are attributed to differences in biological and physiological processes inherent to the individual, rather than conceptualized as a product of socially constructed power relations between groups. Moreover, the power generated through the constitution of these knowledge claims and acts of judgment about health risks are magnified by the mask of scientific neutrality. Thus, by *individualizing* bodies and bodily differences, biomedicine as a regime of power–knowledge at the same time acts to *disindividualize* power, investing it not so much in individual practitioners but in the much more diffuse, constant, and pervasive acts of knowledge construction and production.

For example, essentialist conceptions of health risks and risky behaviors that become tied to and embedded within constructions of race, class, and gender come to comprise constructed notions of normality and pathology, conformity and deviance. When *behaviors* are implicated in producing more frequent adverse outcomes, standards of conformity and deviance are created: "In this sense, epidemiology is inevitably a 'normalizing' science, employing—and reinforcing—unexamined notions of normality to measure and classify deviations from the norm" (Epstein 1996, p. 47). In this fashion, "scientific" standards of normality and pathology are established. These standards act to disindividualize power, uncoupling functions of surveillance and regulation from particular persons in dominant positions and instead locating them within diffuse behavioral norms and processes of self-governance.

Foucault's (1977) metaphor of the Panopticon can thus be seen as central to this interrogation of biomedical knowledge production on disease causation and its uses of race, class, and gender. Foucault argues that the exercise of power within a particular context establishes a site (e.g., the clinic; Foucault 1975) for the "hierarchical observation" of bodies. Achieving greater efficiency of surveillance requires particular types of architecture, specifically the organization of the Panopticon. Panoptic architecture comprises a central tower surrounded by a circular arrangement of cells in which individuals are isolated from one another while remaining completely visible to an observer in the central tower. In fact, ideally, individuals in the cells cannot ascertain whether anyone is actually in the tower; however because the individuals themselves are immediately visible, they conduct themselves *as if* someone is always observing them. This arrangement is therefore structured such that the regulation of individual bodies and conduct no longer depends upon the actual presence or specific identities of dominating agents.

More than a kind of architecture, however, the Panopticon serves as a generalized model of functioning in modern society, a way of rendering discipline increasingly efficient and self-regulatory. Panoptic forms of knowledge and discipline thereby place power relations within the everyday lives of humans by creating normalizing imperatives used to effect the *self*-judgment and *self*-surveillance of individuals and groups. Thus power is automatically "built in" and much more mobile, no longer dependent upon specific relationships between particular individuals: "Power has its principle not so much in a person as in a certain concerted distribution of bodies, surfaces, lights, gazes; in an arrangement whose internal mechanisms produce the relation in which individuals are caught up" (Foucault 1977, p. 202).

The power–knowledge relation that is biomedical knowledge production—in which bodily differences of race, class, and gender are seen as meaningful and significant in their individualistic conceptions—comprises a Panoptic structure, a mechanism that "automatizes and disindividualizes power" (Foucault 1977, p.

202). Foucault's argument here can help explain how even well-meaning providers, politicized and self-aware patients, and advocacy organizations can inadvertently contribute to the continued domination of racialized, classed, and gendered individuals and groups. Because they are already embedded in power–knowledge relations, processes, and architectures of knowledge production, those involved in the delivery and consumption of services based on biomedical knowledge often reinforce imperatives of individual-level accountability for health and the construction of risk categories and risky bodies based on race, class, and gender. Moreover, in order to advance the rationalization of clinical interventions, the atomistic bodies of individuals then become sites for the further production of epidemiologic knowledge. In turn, they are constructed not as the *potential* objects of medical control under the germ theory of disease, but as the *de facto* objects of epidemiologic surveillance under the current assumption that almost all bodies have one or more health risk factors (Armstrong 1995).

However, drawing again on Foucault, it cannot be assumed that epidemiologic projects have inevitably dominating effects on racialized, classed, and gendered bodies. Power, for Foucault, is uniquely conceptualized not as a property that an agent can possess, but as a strategy or a technique to be exercised (Foucault 1977, 1978, 1980). The variable results that power can effect (i.e., the terminal forms that power can take) such as the sovereignty of the state or overall domination, depend not upon the appropriation of power, but upon its disposition, tactical use, and technical functioning. Power takes shape and comes into play in families, groups, institutions, and in other interpersonal and social relationships, yet it is not exactly localized in them. Power is thus something that is produced and "exercised from innumerable points, in the interplay of nonegalitarian and mobile relations" (Foucault 1978, p. 94). It is embedded within and emerges out of practices and the architecture of social relations.

Thus Foucault argues that while power relations can have intentional objectives of domination, it is impossible to say that they are exercised from a permanent subject position of dominance, or that actions that result in broad social cleavages and hierarchies are coordinated (Foucault 1978). The relationships between the rulers and the ruled cannot be characterized as a simple binary opposition; power, knowledges, and discourses cannot be simplistically divided between accepted and excluded, between valorized and rejected. Instead knowledges and discourses not only function as instruments and emerge as effects of power, but they also serve as points of resistance and departure for oppositional strategies. Resistance is *always* immanent in relations of power, Foucault argues, playing the role of adversary, target, or support in power relations.

However, Foucault does assert that domination and hegemony do exist: as power is produced, it sometimes results in points of density or concentration (Foucault 1978, 1980). Power, when exercised in numerous kinds of relationships, can effect cleavages that run through the social body as a whole, forming general lines of force and hegemonic effects that traverse and generate local oppositions. While biomedical knowledge and its constructions of bodily "differences" are certainly contested, the effects of the dominance of the biomedical paradigm on individuals' lived experiences of those differences deemed to be significant to their health are nevertheless material and pervasive. For a further exploration of the processes by which racial, class, and gender hegemonies are constructed and sustained through epidemiologic projects, I now turn to the theories of Omi and Winant (1994) who take race, and its intersections with class and gender, as their points of departure.

EPIDEMIOLOGICAL PROJECTS AS RACIAL, CLASS, AND GENDER FORMATION

Racial formation theory (Omi and Winant 1994) provides a framework for analyzing the constructions and articulations of race within specific sociohistorical contexts and in mutual determination with other axes of difference and hierarchy. Omi and Winant assert that race must be understood as a fundamental dimension of our social organization and cultural meaning. They define race as a concept "which signifies and symbolizes social conflicts and interests by referring to different types of human bodies... selection of these particular human features for purposes of racial signification is always and necessarily a social and historical process" (1994, p. 55)—a process they term *racial formation*. That is, racial formation is a "sociohistorical process by which racial categories are created, inhabited, transformed, and destroyed" (1994, p. 55), and that is realized through *racial projects* in which human bodies and social structures are represented and organized. These racial projects serve as the linkages between structure and representation:

> a racial project is simultaneously an interpretation, representation, or explanation of racial dynamics, and an effort to reorganize and redistribute resources along particular racial lines. Racial projects connect what race *means* in a particular discursive practice and the ways in which both social structures and everyday experiences are racially organized. (1994, p. 56)

In and through racial projects, the meanings of race, while fluid and unstable, effect material and significant consequences. Multiple social practices emerge out of systems of racial classification and racialized social structures; these practices are so numerous and often unconscious to the point of being almost invisible. That is, "race becomes 'common sense'—a way of comprehending, explaining, and acting in the world" (Omi and Winant 1994, p. 60). The production and adherence to this "common sense" then provides the basis for the organization of a racial hegemony. Moreover, race, class, and gender, in Omi and Winant's view, constitute overlapping and intersecting "regions" of hegemony; as such, political conflicts often play out in some or all of these terrains simultaneously.

Biomedical knowledge production about disease causation and distribution is one site where the meanings of race, class, and gender are defined and contested. Much as the U.S. Census Bureau formulates and debates the meanings of racial categories, biomedical and epidemiologic research taking place every day constructs and authorizes particularistic definitions of what race, class, and gender are and what categories are considered legitimate. More specifically, biomedical theories of disease etiology and distribution claim explanatory power by referring to different types of human bodies, and implicitly and explicitly to their internal physiologies. Such theories construct claims that bodies distinguished by race, socioeconomic status, and sex are essentially different. These claims are then used to explain the disproportionate distribution of illness across populations, particularly to legitimate persistent and profound inequalities by race, social class, and gender.

Thus the practice of a biomedical science that routinely incorporates uncritical and individualistic conceptions of race can be seen to constitute a kind of racial project. It provides a ready explanation based on racial/ethnic categories for health and illness outcomes and in turn directs future scientific practices and resources along a line of inquiry that represents race as an individual-level risk factor. As Omi and Winant (1994, p. 3) note, "how one is categorized is far from a merely academic or even personal matter. Such matters as access to employment, housing, or other publically or privately valued goods; social program design and the disbursement of local, state, and federal funds...are directly affected by racial classification and the recognition of 'legitimate' groups." Surely, biomedicine and the enterprise of producing biomedical knowledge constitute another such matter. The biomedicine of race is a major contributor to "a vast web of racial projects [that] mediates between the discursive or representational means in which race is identified and signified on the one hand, and the institutional and organizational forms in which it is routinized and standardized on the other" (1994, p. 60).

Thus science's and particularly biomedicine's understandings of racial differences, their salience, and their employment in research practices all rationalize and legitimate inequities of health outcomes. As such, conventional biomedical knowledge production can be characterized to be a *racist* racial project. Omi and Winant assert that "a racial project can be defined as *racist* if and only if it *creates or reproduces structures of domination based on essentialist categories of race.*... Our definition...focuses...on the 'work' essentializing does for domination, and the 'need' domination displays to essentialize the subordinated" (Omi and Winant 1994, p. 71). While race can and does affect health,[8] the convention of using racial categories to characterize individuals and account for their health status without considering the meanings or social nature of those categories routinizes and legitimates *essentialist* constructions of race as an individualistic attribute.

In an analogous fashion, we can view the constructions of social class and gender and their use in biomedicine as projects—akin to racial projects—that connect discursive meanings of class and gender to institutionalized forms in which those meanings are routinized and standardized.[9] Additionally, many biomedical knowledge production endeavors around class and gender in disease causation and distribution can be defined as *classist* and *sexist* projects because they provide symbolic infrastructures for class and gender domination through essentialized meanings of what it is to belong to a particular class or sex category.

By erasing the social relations of power embodied in race, class, and gender, the essentialism of their biomedical constructions suggests that individual risk assessment sufficiently explains the incidence of disease and patterns of health outcomes. The rational aura of biomedical science cultivates the ideology that any health inequalities that happen to fall along racial, class, or gender lines must be due˜ to individual *biological* or *behavioral* characteristics since their causes are not viewed as *socially* skewed. In short, essentializing practices often characteristic of epidemiologic knowledge production accomplish three things that can contribute to domination: First, they further stabilize the notion that race, class, and gender are features of and dynamics occurring *within individual* bodies. Second, they construct and *naturalize* supposed differences between those categories of bodies. And third, they consequently obscure the *social relations* of power that produce health inequalities. Through such processes and effects, epidemiologic conventions and practices can be seen as resonating with and mutually reinforcing ideological discourses about the place and effects of race, class, and gender that exist within the sciences[10] and permeate contemporary U.S. culture. In the United States, the meanings attached to race, class, and gender—and to specific racial, class, and gender groups—form taken-for-granted sets of beliefs used as tools and justifications for, on the one hand, domination and privilege, and on the other, oppression and exclusion. In these ways, race, class, and gender are *ideologies* used in the exercise of power and in the maintenance of existing structures of privilege and distribution of wealth and resources. In the arena of health and illness, these ideologies structure particular histories of experiences and life conditions that in turn shape exposure to injury and disease and abilities to access resources.

As for the possibilities for changing the meanings accorded to bodily "differences" and the hegemonies in which they are institutionalized, Omi and Winant argue that the trajectory and shape of contemporary racial politics are largely determined by the pattern of conflict and accommodation between the state and racially based social movements. Racial meanings and hegemony are disrupted as social movements and the state move through periods of unstable equilibria, crisis, absorption, insulation, and adjustment (Omi and Winant 1994). More generally, Omi and Winant suggest that shifts in the relationships between social movements and the state contribute to the construction of what race, class, and gender *are*, what their meanings might be, and the consequences they will have on resources, policies, and the larger social order.

Certainly biomedicine as a racial, class, and gender project are affected by the dynamics between the state and social movements. Biomedical science occupies a complex position with regard to the state and social movements: biomedical knowledge production is heavily financed and even directly undertaken by federal, state, and local government entities, and increasingly, the public and social movements play a role in shaping what the research priorities should be. In another example, state recommendations on racial categorization, and its agenda regarding questions of racial, socioeconomic, and gender disparities dictate to a great extent what research is done and how it should be conducted. Furthermore, conventional biomedical knowledges have proven beneficial to organized social movements: in helping to build a case for the need for social resources to address racial, class, and gender inequalities in health status, treatment, and access, they provide the basis for activism and a yardstick by which social change can be measured. Thus social movements do play prominent roles in mediating racial, class, and gender projects and do have critical impacts on biomedical knowledge production (e.g., Epstein 1996).

At the same time, however, the nature of knowledge production about matters of health, illness, and bodily difference is such that much of it takes place almost "outside of," and even "prior to," conspicuously political contestations between the state and social movements. Individuals evaluate medical information and knowledge claims within the context of their own lived experiences of illness and attempt to incorporate, revise, or reject them accordingly. Such everyday struggles over the veracity and applicability of biomedical science on race, class, and gender to individual lives and collective experiences have the potential to shape the politics of biomedical research. That is, these kinds of dynamics have profound impacts on biomedical knowledge production as a project of racial, class, and gender formation, and on the maintenance or contestation of biomedical, racial, class, and gender hegemonies. Theories, ideas, ideologies, assumptions, and conclusions about the relationships between disease and race, class, and gender are ubiquitous, frustrating attempts to separate biomedical and "expert" from so-called "lay" knowledge production.

In sum, the construction of categories and meanings of race, class, and gender in relation to health and illness does not occur solely in biomedical settings and organized social movements, nor is it necessarily immediately involved in the allocation of resources. While there certainly is contestation over health-related funds and services, there is also contestation over those knowledge claims and constructions that provide ideological legitimacy to hierarchical systems and social structures through which health inequalities are produced. Indeed, such contestations provide the "raw material" from which more organized social movements can emerge. Thus a full understanding of racial, class, and gender hierarchies must incorporate an explicit examination of the *construction of knowledges*—the practices and sites of their production, the extent to which (in this case) biomedical knowledge is "taken up," transformed, and contested by "lay" people (see e.g., Cooter and Pumfrey 1994; Epstein 1995; Kroll-Smith and Floyd 1997; Wynne 1996)—and their links to power and social order.

CONCLUSION: THE CONTINUED MYSTIFICATION OF THE SOCIAL PRODUCTION OF HEALTH

In operationalizing the multifactorial model of disease causation, contemporary epidemiology focuses particularly on specific personal behavioral risk factors, imparting primary significance to individual attributes in the explanation of health inequalities. "By concentrating on these specific and presumed free-range individual behaviors, we thereby pay less attention to the underlying social–historical influences on behavioral choices, patterns, and population health" (McMichael 1995, pp. 633-634). As such, epidemiology contributes to the construction of standards of normality and pathology in which the individual and her/his behaviors are targeted as the objects of control. A preoccupation with the intricate and complex modeling of risk factors and behaviors has superseded epistemological concerns with understanding their origins and implications for health policies and practices (Krieger 1994). This almost exclusive focus makes even the underlying multifactorial model of disease causation implicit—treated as a given—in the practice of epidemiology (Krieger 1994), rather than a site around which the dynamics and production of stratified health can be explicitly theorized.

Disease causation and the social production of health inequalities are extremely complex and messy; they involve transhistorical, interactive, and perhaps currently unmeasurable social processes and relations (Krieger and Bassett 1993). Yet, epidemiology and the deployment of epidemiologic claims in clinical practices sidestep such complexity in efforts to yield pragmatic and simplistic policies that do little to challenge the status quo. As Wynne (1996, pp. 20-21) notes, "It is not…that scientific knowledge merely *omits* social dimensions that ordinary people incorporate in their evaluations and assessments. It is that scientific knowledge tacitly imports and imposes particular and problematic versions of social relationships and identities." Because the use of essentialist constructions to inform clinical and health promotion practices emerges from a seemingly objective and rational scientific methodology, the ultimate outcome is the further reification—through scientific legitimation—of racial, class, and gender hierarchy. This reductionist mode of thinking and individualistic constructions of race, class, and gender have proved to be quite mobile and resilient, transgressing presumed boundaries between science, politics, and society, and shaping, as well as being shaped by, the experiences of well-being or ill-health of individuals.

This does not mean that we should give up the epidemiologic endeavor and the study of health inequalities through the use of racial, socioeconomic, and gender categories. Race, class, and gender clearly *do* matter in the distribution of health and illness, and the embodied, material health effects that are the consequences of constructions of difference challenge the notion that awareness of racial, class, and gender differences are but psychological and attitudinal "problems." To stick to the argument that race, class, and gender have no place in biomedical science or, conversely, to give them a place but avoid an explicit discussion of what this place is about, are equally dangerous directions. These positions are related to the fallacious belief that we as a society can be and should be blind to racial, class, and gender differences—that they represent anachronistic concerns that no longer matter—or, conversely, that their meanings are self-evident and unproblematic.

Instead, I assert that a clearer comprehension of the parts that race, class, and gender play in matters of health and illness requires that we understand biomedical knowledge production as a participant in the social construction of race, class, and gender, and their linkage to institutional practices and processes. That is, far from being an "objective" and "neutral observer," biomedicine is an active participant in the construction of difference. Investigating epidemiologic knowledge in light of Foucauldian and racial formation theories makes this clear. We can then be in a position to raise questions about biomedicine's dominance over the "definition of the situation" around racial, class, and gender inequalities in health, and the potential implications of accepting its frames.

I believe such questions are especially important to raise given our current historical and social context. We live in a time and place that is marked by what many have characterized as a backlash against (or at the very least, a regressive slide from) the advances made since the women's, civil rights, and anti-poverty movements throughout this century. Significant changes in the health care landscape have emerged, including the rise of risk discourses and genomics, the expansion of managed care, rollbacks in federal and state welfare and social programs, and a swelling chorus of health promotion and prevention rhetorics and activities. All of these changes, it would seem, represent a devolution in health care, services, and discourses (Estes and Linkins 1997), a downshift from governmental and social responsibility for the distribution of health and illness, to a demand for heightened individual accountability.

At the same time, however, numerous contradictions pervade this landscape and continually compel our attention. For example, we are increasingly dependent upon science in general and biomedicine in particular, yet simultaneously we grow ever more aware of their limitations and fragility (e.g., Beck 1992; Epstein 1996; Giddens 1991; Irwin and Wynne 1996). Multiple refutations of the scientific basis for the inequality of the races (e.g., UNESCO 1952)[11] have been articulated, yet measures of race, as well as of class and gender, remain ubiquitous and meaningful elements of biomedical knowledge production and discourse (Osborne and Feit 1992). And finally, human pain, destitution, and suffering continue to persist, amidst enormous wealth and technological and scientific activity. Such contradictions and enduring problems seem to demand that we examine anew the social meanings we have invested in biomedical knowledge and the position that it occupies in our individual and collective affairs.

To conclude, by focusing on health risks to explain behaviors and interactions and by not looking "upstream" (McKinlay 1974/1997) to social dynamics of power and hierarchies of knowledge, biomedical researchers ignore that risks are *embedded in* and *shaped by* structural conditions of existence. Race, class, and gender are thus seen not as relational constructs but as individualistic attributes, as located solely *in* the biological bodies of individuals rather than also in the social spaces *between* them. Race, class and gender must be reconceptualized as social relations of power that are fundamentally causal to the etiology and distribution of health and illness. These relations of power produce health inequalities through a variety of intermediate mechanisms and proximal causes that are only incidental to the overall process and thus may shift across historical and social contexts. In order to eliminate racial, class, and gender inequalities in health, we must directly address the more fundamental relational processes and practices that link racism, classism, sexism, and health.

ACKNOWLEDGMENTS

A portion of this paper was presented at the 1998 Annual Meeting of the American Sociological Association. I am grateful to Jennie Jacobs Kronenfeld for her editorial guidance. Special thanks also go to Adele E. Clarke, Howard Pinderhughes, Evelyn Nakano Glenn, Jennifer Ruth Fosket, Laura Mamo, Teresa Scherzer, and Jennifer R. Fishman for their support and comments on earlier versions of this paper.

NOTES

1.   In this paper, I use the term "biomedicine" to refer to theories and practices of healing that belong to the tradition of science, and specifically to allopathic medicine, in the post-Enlightenment Western world.

2.   Although, interestingly, the *measurement* of sex is not, at this historical moment, seen to be problematic or even at issue; that is, the assumption that there are two mutually exclusive categories of sex/gender goes, for the most part, unquestioned (see, for example, Fausto-Sterling 1993, Kessler 1990).

3.   Host factors include, for example, immunologic status, genetic background, socioeconomic level, hygienic practices, behavioral patterns, age, and the presence of co-existing disease.

4.   For example, the Health Risk Appraisal (HRA) was developed out of the work of Robbins and Hall (1970) as a method to help physicians practice preventive medicine by focusing prospectively on the avoidance of premature mortality. The objective of this intervention is to reduce the burden of premature death ostensibly by providing the information and motivation for individuals to stop or modify "risky" behaviors.

5.   Also called "the new genetics," genomics refers to "a body of knowledge and techniques arising out of the discovery of recombinant DNA in the 1970s. It principally involves research into the genetic components to a range of disease, illness, and behavior and its application in the clinic in the form of testing, screening, and treatment" (Kerr, Cunningham-Burley, and Amos 1998, p. 194).

6.   The social construction of race, gender, and class is "always already" a dialogue among competing dominant and alternative definitions; the construction of "difference" as a symbol of commu-

nity, history, identity, and agency has been an undeniably powerful force affecting social movements, policies, and knowledges around health and illness. However, since my objective is to critique dominant biomedical and epidemiological knowledge production practices, I focus on the individualizing effects of biomedical constructions and their oppressive implications.

7.   For example, see Chase (1998), Fausto-Sterling (1993), and Kessler (1990).

8.   Indeed, taking from my argument the implication that race, class, and gender do not and should not matter in health and illness would be committing precisely the kind of "ideological trick" (Smith 1990) or "god trick" (Haraway 1997) that I am arguing against. It is not an issue of *whether* or not race, class, and gender matter; they *do* matter. What is at issue here is *how* race, class, and gender are conventionally conceptualized and operationalized in biomedical and epidemiologic research, and the kinds of interpretations of research results that are commonly proposed.

9.   While I am usually hesitant to apply theoretical frameworks developed for one dimension of inequality and power to others, I believe this is one instance in which such an application can be argued to be illuminating in thinking about gender and class. In addition, Omi and Winant, as well as the other theorists on race (e.g., Almaguer 1994; Davis 1981; Gilroy 1987; Hall 1996a, 1996b; hooks 1984; Lowe 1996), explicitly include dimensions of gender and class, and sometimes sexuality, in their analyses of racial formation. These scholars share a common orientation that racial formation often works through gender and class formation, that gender formation depends upon meanings and institutions of race and class, and that class formation can only be analyzed in relation to racial and gender formation processes.

10.   Although, the social sciences have been and continue to be one of the few sites where atomism and individualization are challenged.

11.   In 1950 and 1951, the United Nations Educational, Scientific, and Cultural Organization (UNESCO) released statements strongly refuting a scientific basis for the inequality of races (Haraway 1989, p. 197-203; UNESCO 1952). Yet at the same time, UNESCO seemingly reaffirmed the genetic basis of race by defining it as one of a group of populations that differ due to "evolutionary factors of differentiation such as isolation, the drift and random fixation of the material particles that control heredity (the genes), changes in the structure of these particles, hybridization, and natural selection" (UNESCO 1952, p. 98).

REFERENCES

Almaguer, T. 1994. *Racial Fault Lines: The Historical Origins of White Supremacy in California.* Berkeley: University of California Press.

Armstrong, D. 1995. "The rise of surveillance medicine." *Sociology of Health and Illness* 17: 393-404.

Bayne-Smith, M. 1996. "Health and women of color: A contextual overview." Pp. 1-42 in *Race, Gender, and Health,* edited by M. Bayne-Smith. Thousand Oaks, CA: Sage Publications.

Beck, U. 1992. *Risk Society: Towards a New Modernity.* Beverly Hills, CA: Sage Publications.

Becker, G. and R. Nachtigall. 1994. "'Born to be a mother:' The cultural construction of risk in infertility treatment in the U.S." *Social Science and Medicine* 39: 507-518.

Chase, C. 1998. "Hermaphrodites with attitude: Mapping the emergence of intersex political activism." *GLQ-A Journal of Lesbian and Gay Studies* 4: 189-211.

Cooper, R.S. 1994. "A case study in the use of race and ethnicity in public health surveillance." *Public Health Reports* 109: 46-52.

Cooter, R. and S. Pumfrey. 1994. "Separate spheres and public places: Reflections on the history of science popularization and science in popular culture." *History of Science* 32: 237-67.

Davis, A.Y. 1981. *Women, Race, and Class.* New York: Random House.

Davis-Floyd, R.E. 1992. *Birth as an American Rite of Passage.* Berkeley: University of California Press.

Epstein, S. 1995. "The construction of lay expertise: AIDS activism and the forging of credibility in the reform of clinical trials." *Science, Technology and Human Values* 20: 408-37.

_____. 1996. *Impure Science: AIDS, Activism, and the Politics of Knowledge*. Berkeley, CA: University of California Press.

Estes, C.L. and K.W. Linkins. 1997. "Devolution and aging policy: Racing to the bottom of long term care?" *International Journal of Health Services* 27: 427-442.

Evans, A.S. 1978. "Causation and disease: A chronological journey." *American Journal of Epidemiology* 108: 249-58.

Fausto-Sterling, A. 1993. "The five sexes: Why men and women are not enough." *The Sciences* 33: 2-25.

Foucault, M. 1975. *Birth of the Clinic*. Translated by AM Sheridan Smith. New York: Vintage Books.

_____. 1977. *Discipline and Punish: The Birth of the Prison*. Translated by Alan Sheridan. New York: Vintage Books.

_____. 1978. *The History of Sexuality, Volume I: An Introduction*. Translated by Robert Hurley. New York: Vintage Books.

_____. 1980. *Power/Knowledge: Selected Interviews and Other Writings 1972-1977*, Edited by C. Gordon. New York: Pantheon Books.

Giddens, A. 1991. *Modernity and Self-Identity: Self and Society in the Late Modern Age*. Stanford: Stanford University Press.

Gilroy, P. 1987. *There Ain't No Black in the Union Jack: The Cultural Politics of Race and Nation*. London: Hutchinson.

Ginzburg, H.M. 1986. "Use and misuse of epidemiological data in the courtroom: Defining the limits of inferential and particularistic evidence in mass tort litigation." *American Journal of Law in Medicine* XII: 423-439.

Guralnick, J.M. and S.G. Leveille. 1997. "Annotation: Race, ethnicity, and health outcomes—unraveling the mediating role of socioeconomic status." *American Journal of Public Health* 87: 728-730.

Hahn, R.A. and D.F. Stroup. 1994. "Race and ethnicity in public health surveillance: Criteria for the scientific use of social categories." *Public Health Reports* 109: 7-15.

Hall, S. 1996a. "New ethnicities." Pp. 441-449 in *Stuart Hall: Critical Dialogues in Cultural Studies*, edited by D. Morley and K.-H. Chen. New York: Routledge.

_____. 1996b. "What is this 'black' in black popular culture?" Pp. 465-75 in *Stuart Hall: Critical Dialogues in Cultural Studies*, edited by D. Morley and K.-H. Chen. New York: Routledge.

Haraway, D. 1989. *Primate Visions: Gender, Race and Nature in the World of Modern Science*. New York: Routledge.

_____. 1997. *Modest_Witness@Second_Millenium.FemaleMan©_Meets_Oncomouse™: Feminism and Technoscience*. New York: Routledge.

hooks, b. 1984. *Feminist Theory: From Margin to Center*. Boston: South End Press.

Irwin, A. and B. Wynne. 1996. "Introduction." Pp. 1-18 in *Misunderstanding Science? The Public Reconstruction of Science and Technology*, edited by A. Irwin and B. Wynne. Cambridge, England: Cambridge University Press.

Kerr, A., S. Cunningham-Burley, and A. Amos. 1998. "Eugenics and the new genetics in Britain: Examining contemporary professionals' accounts." *Science, Technology, & Human Values* 23: 175-198.

Kessler, S.J. 1990. "The medical construction of gender: Case management of intersexed infants." *Signs: Journal of Women in Culture and Society* 16: 3-26.

Krieger, N. 1994. "Epidemiology and the web of causation: Has anyone seen the spider?" *Social Science and Medicine* 39: 887-903.

Krieger, N. and M. Bassett. 1993. "The health of black folk: Disease, class, and ideology in science." Pp. 161-69 in *The "Racial" Economy of Science: Toward a Democratic Future*, edited by S. Harding. Bloomington, IN: Indiana University Press.

Kroll-Smith, S. and H.H. Floyd. 1997. *Bodies in Protest: Environmental Illness and the Struggle Over Medical Knowledge*. New York: New York University Press.

LaVeist, T.A. 1996. "Why we should continue to study race...but do a better job: An essay on race, racism, and health." *Ethnicity and Disease* 6: 21-29.

Lowe, L. 1996. *Immigrant Acts: On Asian American Cultural Politics*. Durham, NC: Duke University Press.

Lupton, D. 1994. *Medicine as Culture: Illness, Disease and the Body in Western Societies*. Thousand Oaks, CA: Sage Publications.

MacMahon, B., T. F. Pugh, and J. Ipsen. 1960. *Epidemiologic Methods*. Boston: Little, Brown & Co.

Martin, E. 1987. *The Woman in the Body: A Cultural Analysis of Reproduction*. Boston: Beacon Press.

McKenney, N.R. and C.E. Bennett. 1994. "Issues regarding data on race and ethnicity: The Census Bureau experience." *Public Health Reports* 109: 16-25.

McKinlay, J.B. 1974/1997. "A case for refocussing upstream: The political economy of illness." Pp. 519-33 in *The Sociology of Health and Illness: Critical Perspectives*, edited by P. Conrad. New York: St. Martin's Press.

McMichael, A.J. 1995. "The health of persons, populations, and planets: Epidemiology comes full circle." *Epidemiology* 6: 633-636.

Nickens, H.W. 1995. "The role of race/ethnicity and social class in minority health status." *Health Services Research* 30: 151-162.

Oliver, M.L. and T.M. Shapiro. 1995. *Black Wealth/White Wealth: A New Perspective on Racial Inequality*. New York: Routledge.

Omi, M. and H. Winant. 1994. *Racial Formation in the United States: From the 1960s to the 1990s*. New York: Routledge.

Osborne, N.G. and M.D. Feit. 1992. "The use of race in medical research." *Journal of the American Medical Association* 267: 275-279.

Pearce, N. 1996. "Traditional epidemiology, modern epidemiology, and public health." *American Journal of Public Health* 86: 678-683.

Robbins, L. and J. Hall. 1970. *How to Practice Prospective Medicine*. Indianapolis, IN: Methodist Hospital of Indiana.

Roberts, D. 1997. *Killing the Black Body: Race, Reproduction, and the Meaning of Liberty*. New York: Pantheon Books.

Schulman, K.A., E. Rubenstein, F.D. Chesley, and J.M. Eisenberg. 1995. "The roles of race and socioeconomic factors in health services research." *Health Services Research* 30: 179-195.

Smith, D. 1990. *The Conceptual Practices of Power: A Feminist Sociology of Knowledge*. Boston: Northeastern University Press.

Susser, M. 1985. "Epidemiology in the United States after World War II: The evolution of technique." *Epidemiology Review* 7: 147-177.

Susser, M. and E. Susser. 1996. "Choosing a future for epidemiology: II. from black box to Chinese boxes and eco-epidemiology." *American Journal of Public Health* 86: 674-677.

U.S. Bureau of the Census. 1993. "Money Income of Households, Families, and Persons in the United States: 1992." U.S. Government Printing Office, Washington DC.

UNESCO. 1952. "The Race Concept: Results of an Inquiry." UNESCO, Paris.

Villemez, W.J. 1980. "Race, class, and neighborhood: Differences in the residential return on individual resources." *Social Forces* 59: 414-430.

Williams, D.R. 1997. "Race and health: Basic questions, emerging directions." *Annals of Epidemiology* 7: 322-333.

Williams, D.R. and C. Collins. 1995. "U.S. socioeconomic and racial differences in health: Patterns and explanations." *Annual Review of Sociology* 21: 349-386.

Wynne, B. 1996. "Misunderstood misunderstandings: Social identities and public uptake of science." Pp. 19-46 in *Misunderstanding Science? The Public Reconstruction of Science and Technology*, edited by A. Irwin and B. Wynne. Cambridge, MA: Cambridge University Press.

PART IV

RESTRUCTURING OF CARE

PREDICTORS OF USE OF TELEMEDICINE FOR DIFFERING MEDICAL CONDITIONS

William Alex McIntosh, Letitia T. Alston,
John R. Booher, Dianne Sykes, and Clasina B. Segura

ABSTRACT

Telemedicine has been touted as a solution to the problems of access to health care experienced by rural and other isolated populations. Few studies exist, however, that explain the differential use of telemedicine by patients. Patient utilization of this technology is likely to be predicted by both those factors that affect the adoption of new ideas, on the one hand, and those factors that affect the utilization of health care, on the other. These include (1) propensity factors such as education and ethnicity; (2) enabling factors such as income and health insurance, and (3) need characteristics such as the severity of the illness as well as factors that impact (4) accessibility. Data were collected from samples of patients who had experienced telemedicine in one of two rural locations in Texas. A second sample of patients who had not experienced telemedicine was drawn from these same locations for comparative purposes. Models were developed to differentiate among patients who would elect to

Research in the Sociology of Health Care, Volume 17, pages 199-213.
Copyright © 2000 by JAI Press Inc.
All rights of reproduction in any form reserved.
ISBN: 0-7623-0644-0

use telemedicine for each of six hypothetical medical conditions versus the alternatives of consulting their local physician without the presence of telemedicine or travel to see a specialist. Separate models were developed for those patients who had previous experience with telemedicine and for those without such previous experience. Prior experience was associated with a greater willingness to use telemedicine for all but the most serious of medical conditions. Predisposition factors had the greatest impact on the more serious conditions (e.g., problem pregnancy and cancer), and enabling conditions as well as access factors were more likely to affect the less serious conditions (e.g., cough and rash).

INTRODUCTION

Access to specialty care remains a significant concern to many rural and small town residents. Most such communities lack a sufficiently large patient base to support the placement of specialists locally. Many rural residents thus face costly travel in order to access specialty care. However, residents in rural areas and small towns may lack the necessary transportation, time, or monetary resources required to go to a larger community to obtain specialty care.

One solution to this problem is the implementation of a telemedical system that permits patients access to specialists via two-way interactive consultation either in the office of their personal physician or at the local hospital/clinic. There are approximately 80 active telemedicine projects in the U.S.: 47 of these conduct at least 50 or more consultations per year (Grigsby and Allen 1997; Mintzer et al. 1997). A number of studies of patient satisfaction with telemedicine and of cost savings have now been conducted. However, there have been no studies of patient willingness to adopt telemedicine. Using a framework based on health care utilization and adoption of innovation models, the present study compares patients who have experienced telemedicine to those who have not to determine what kinds of persons would more likely use it under a variety of circumstances.

Literature Review

Telemedicine is a variation on traditional health services, and participation in telemedicine can be viewed as a form of health care utilization. Models of health care utilization are, therefore, relevant to the understanding of patient use of telemedicine. It also represents a new technology for patients, and its use, therefore, incorporates an element of adoption by patients as an accepted way of seeking care. Thus, adoption models are appropriate as well. Because the two kinds of models utilize many of the same kinds of concepts, combining the health care utilization and adoption of innovation models does not require a logical stretch.

Health care utilization studies have led to models involving demographic, social structural, social psychological, and geographic variables (Wolinsky 1988).

Many of these variables are found in the Aday/Anderson model of health care utilization (Aday and Awe 1997). This model argues that the use of health care services depends on *propensity,* ability or *enabling* characteristics, *need* at the patient level, and *availability* at the community level. Individuals who are predisposed (propensity) to make use of health care, who have the resources needed to access it, and who have some need (e.g., an illness, or a desire to remain healthy) are more likely to use such services than those who lack these characteristics.

Propensity involves those characteristics that would lead a person to seek care. Some of these characteristics are social psychological in nature; for example, locus of control. Others have to do with potential need and involve demographic characteristics such as age, gender, marital status, and family size and composition. The very young and the very old, for example, generally require more health care than others. Women tend to take fewer risks and thus engage in not only more preventive health behaviors than men, but are also more prone to seek medical advice. Occupation is associated with differing lifestyles, which include use of medical services.

Ethnicity has been an important predisposing variable in these models. Early research showed that members of minority groups were consistently shown to use fewer health care services than the majority group. More recent work has demonstrated, however, that minorities use health care as frequently as the majority (Geertsen 1997). Gochman (1997) observes that while rates of use differ little at present, ethnic groups have differing combinations of enabling and predisposing factors associated with such utilization. He argues that these differences reflect continued cultural differentiation among ethnic groups.

Enabling characteristics include the resources needed to access existing facilities. Such resources include time, transportation, money, and information. In order to use a service, a person has to have some awareness that the service exists, be able to get to that service, and pay for its use. Income and health insurance are, therefore, frequently discussed as enabling conditions. The health care utilization model assumes that, in order to use services, individuals must be aware that they exist. These models, thus, do not take into account nonutilization due to lack of information. Such a lack of awareness might well be a factor if the services offered differ from those normally available to the individual.

Need for medical care is considered the most important predictor and is usually measured in terms of number/severity of symptoms, disability days, or the individual's subjective evaluation of health. The early Anderson model emphasized patient demand for health care; the later Aday/Anderson formulation included issues of *supply* or availability: the type and amount of health care services available in the community. Closely related to this concept is distance to health care services, an important issue for persons living in smaller communities or rural areas or in large cities without public transportation. The greater the distance to health services, the less likely the service is to be utilized (Ricketts and Savitz,

1994). Climate and terrain may also make travel more difficult, adding to the salience of distance.

Adoption of technology models have tended to emphasize demand rather than supply. Demand characteristics include social structural and social psychological variables that are similar, in many ways, to those used in the health care utilization models. For example, age is an important factor in *propensity* to adopt. Adopting new technology involves some degree of risk in that the technology might not generate the type or amount of outcomes anticipated. Younger persons are more likely to take such risks and thus are more likely to adopt innovations. This is true of both physicians and farmers (Fenwick and Schwartz 1994; Rogers, 1995).

Enabling factors have been found to be more important than propensity in many adoption studies. Income and education are the two most frequently cited variables in this regard. Those persons with more income can afford to take more risks than those persons with less money. Information on an innovation is another enabling factor. Physicians and farmers are more likely to adopt new technology if they have complete information. Other researchers have thus focused on the type and amount of information regarding the technology and its availability. Education is associated with more knowledge of innovations and greater seeking of information regarding changing technology. Such knowledge begins with an awareness that the technology exists.

Need is rarely discussed in the adoption literature. Although the perception of the innovation as offering an advantage is important in the attractiveness of the innovation, the need for the innovation is logically antecedent to the adoption process. However, need has received little attention in the research literature. Nevertheless, Hassinger (1959), among others, has argued that people are not attentive to information about an innovation unless there is a perceived need for it. Need is, therefore, an implicit element of this tradition. Finally, Brown (1981) has argued for a "supply side" emphasis in adoption-diffusion studies. He observes that the distance between the potential adopter and the sources of the innovation affects the rate of adoption: that is, the greater the distance between the adopter and the innovation, the greater the amount of time it takes for adoption to occur. Thus, both the health care utilization and adoption models contain supply constraints.

Studies of adoption have also taken into account the degree to which an innovation can be easily tried out. The more *"trialable"*, the more likely the innovation will be tested and then adopted. Those who tried out a technology or service are more likely to perceive its strengths and limitations and this affects willingness to use the technology or service again.

Our study develops prediction models of the adoption of telemedicine for various medical conditions that are based on the Aday/Anderson health care utilization model and the classical adoption–diffusion literature. These models differentiate among preferences to adopt telemedicine, travel to see a specialist, or see the local health care provider without using telemedicine. We also develop models which distinguish between those who have previously experienced telemedicine and those with no previous experience.

Methodology

The data were collected as part of a larger project designed to study community adoption of telemedicine. Patient adoption was measured by satisfaction with the experience, and willingness to do it again is obviously an important component of this. We obtained data from two major telemedicine projects in Texas. The first took place in central Texas over a two-year period in the early 1990s. The second project began in western Texas in September 1990 and has continued since. The two areas vary a great deal. The central Texas project included patient participants from the county where the telemedicine equipment was located as well as from six surrounding counties. These counties are located in south central Texas and are about 100 miles northeast of Houston. The counties are classified as "sparsely settled rural, with some urban" using Bluestone's county classification scheme (Bluestone and Daberkow 1985). Agriculture has been the primary economic activity, although the area has undergone both an oil boom and oil bust in the past 10 years. Almost all of the counties in this region have at least one community hospital, but few have much in the way of specialty care. Access to specialists is obtained largely by driving to a large city, approximately an hour's drive for most residents. The specialists who participated in the telemedicine project for this region were located in this city.

The western Texas site involves six counties, two of which have telemedical technology installed. These counties are contiguous with the Mexican border in the Trans-Pecos region of the state. It should also be noted that western Texas counties are significantly larger in geographical area than central Texas counties. The area is well-over 100 miles from El Paso (to the west) and Midland-Odessa (to the north). Access to specialty care for residents is much more difficult than it is for those in the central Texas area. The counties are sparsely populated, with densities low enough for the counties to be classified as "frontier" in nature. The area is primarily agricultural in its economy, but some tourism is also important, given the proximity of a national park. The specialists are primarily located in a large city, several hundred miles away from the western Texas site.

Both areas have high rates of poverty relative to other parts of the state. The western Texas site is characterized by a high percentage of Hispanics and a low percentage of Blacks; the central Texas area is somewhat similar to the rest of the state in terms of ethnic composition with 14% Blacks, but only 6% Hispanics.

Sample

The patient–user samples were drawn from lists of patients who had used telemedicine as part of a consultation with a specialist. All patients who had participated signed a consent form prior to their consultations that gave permission for researchers to record and analyze data obtained from the patient and from the consultation. At the central Texas site, equipment was located at a community hospital, a mental health clinic, a renal dialysis clinic, and a juvenile detention center. The bulk of the consultations took place at the renal dialysis center, and this center was also the only place where we were able to access records of patients who had used telemedicine. Nearly 100 persons had participated in telemedicine at the renal dialysis clinic. However, because of the nature of their aliments and their age, many of these patients had either moved away or died since participating in the project. This resulted in a sample of only six telemedicine users from central Texas. In contrast, the community hospital has been the center of activity in the western Texas project. This hospital has retained a fairly complete record of the telemedical consultations done there. From this project's inception to June 1997, over 300 persons had participated in one or more teleconsultations to access dermatologists, cardiologists, surgeons, neonatologists, orthopedists, and other specialists. We were able to interview 78 of these individuals. The records of the teleconsultations included an indication of the county in which the patient resided.

Using the proportion of users from each county as sample weights, random digit dialing was used to contact 105 central Texas and 106 western Texas nontelemedicine users. The total sample size of users and nonusers was 305.

Measurement

Dependent Variables

Respondents were presented with six scenarios and asked a series of questions regarding how they might behave under each. The hypothetical situations included pregnancy, difficult delivery, cancer symptoms, a child's ear infection, a persistent rash, and a persistent cough. Respondents were asked how they would respond to each scenario. Their options were to do nothing, see their local physician, travel to see a specialist, or see a specialist via telemedicine (with their physician present). We developed two sets of models from these responses in order to contrast those who would use telemedicine rather than travel to see a specialist and those who would use telemedicine as opposed to receiving care solely from their local doctor.

Independent Variables

We measured predisposing factors such as age, gender, number of children in the household, ethnicity, ability to speak English, and regional location. In addition, respondents were asked to speculate regarding the desirability of interacting with specialists via telemedicine compared with traveling to see specialists.

Responses to this open-ended question included a preference for face-to-face contact and mistrust of technology as reasons for selecting traveling over telemedicine; convenience, lack of transportation, and trust of technology were reasons given for the telemedicine preference. We also measured the distance a person was willing to travel to obtain medical care, the distance to the nearest hospital, and whether the respondent had a regular source of medical care. These latter measures reflected not only predisposing but also enabling factors. Additional enabling factors included education and income, which are relevant to both the receipt of information about technology and the ability to take advantage of it. Respondents were also asked whether they had heard of their local telemedicine project. In addition, access to transportation and past experiences with travel to medical care are considered to be enabling factors. Also, the distance to the nearest hospital reflects the distance required to travel to access telemedicine; the distance to a hospital with specialty care represents one of the costs associated with accessing a specialist for a face-to-face consultation. Because the two areas studied differed so greatly in terms of culture and geography, we created interaction terms between location and the variables just discussed.

We also obtained a measure of health risk taking, an indicator of propensity, by summing the answers to the hypothetical questions involving the distinction between seeking help versus waiting with regard to six hypothetical health conditions. The six hypothetical health conditions presented to the respondents were created to represent a continuum of seriousness. These conditions ranged from emotionally charged (pregnancy) and life threatening (cancer) to conditions that are annoying but not usually considered serious (rash). That these conditions represent a crude continuum of perceived seriousness is substantiated by answers to the question of whether or not the respondent would "seek care immediately or wait" for each of the conditions. More respondents replied that they would seek help immediately for conditions such as cancer than in the case of conditions such as a rash. The differential seriousness in the conditions reflects differences in need for health care.

We used logistic regression to create prediction models for each hypothetical situation. We ran separate models for those who had some exposure to telemedicine (users) and those who had not (nonusers).

Findings

Drawing on the overall sample of nonusers, we began by comparing the percentages of respondents who expressed a preference for using the telemedicine system with their physician for consultation with a specialist rather than consultation with their physicians alone (see Table 1).

There is no clear pattern of willingness to choose a telemedical consultation with a specialist and the local provider over consultation with the local physician without alone across the six conditions. In all but two of the cases (cancer in

Table 1. Would Use Telemedicine in Preference to
Local Physician Alone (in percent) Nonuser Sample

	West Texas	Central Texas
Pregnancy	14	0
Cancer	34	55
Earache	53	34
Cough	34	7
Rash	29	10

central Texas and serious earache in a child in western Texas), consultation with the respondent's physician is preferred by the majority of respondents (see Table 1). Nevertheless, substantial numbers of respondents indicated that they would use telemedicine in preference to more traditional kinds of health service. When data were examined for variables associated with these choices, the following patterns emerged.

Nonusers

The variables found most important in the choice to see a specialist were (1) a general preference for traveling to see specialists rather than use technology such as telemedicine, (2) greater proximity to the location of specialists, and (3) a preference for personal contact with the physician (see Table 2). In contrast, the choice to remain in the local community and use telemedicine was associated with a preference for the convenience of telemedicine, lack of transportation, and prior experience with traveling long distances to see specialists. Lack of facility in English was also a factor. For the least serious condition (rash), a past reliance on the local physician was a key factor in making the choice to stay. In terms of the adoption model proposed above, propensity (willingness to travel), enabling factors (transportation), and access factors (distance from specialists) were all important in this choice.

The finding that male status was associated with a greater preference for travel to see a specialist in the hypothetical case of pregnancy may reflect less familiarity with pregnancy among males and, because of this uncertainty, its elevation to a more serious condition.

An interaction between location and distance was found. In the case of pregnancy, whether or not respondents in western Texas had transportation, they were more likely to choose the telemedicine alternative than respondents were from central Texas. This may well reflect the much greater distances that need to be covered in western Texas. Western Texas respondents indicated that some had traveled as far as 133.5 miles to see a specialist. While they lived an average of 23 miles from the nearest hospital, they lived almost 152 miles from the next nearest hospital, which

Table 2. Logistic Regression of Use of Telemedicine Versus Travel

| | Patients Seek Treatment Outside | | |
Pregnancy	Parameter Estimate	Standardized Coefficient	Odds Ratio
West Texas X Prefer Drive	−2.539**	−.530	.079
Convenience	2.233*	.522	9.327
Prefer Personal Contact	−2.378*	−.562	.130
Distance Traveled to Specialist	.008**	.689	1.008
Male	−3.231**	−.839	.040
$X^2=29.67$***; Pseudo $R^2=.682$ (n=59)			
Cancer			
West Texas X Prefer Drive	2.719***	.621	15.165
No Transportation	2.309**	.428	10.042
Time in Community	−.025**	−.308	.975
White	.976**	.453	2.652
$X^2=34.982$***; Pseudo $R^2=.361$ (n=124)			
Earache			
West Texas X Child Present	−1.580**	−.279	.209
Spanish Speaker	−2.099**	−.137	.123
$X^2=5.164$***; Pseudo $R^2=.078$ (n=87)			
Rash			
Medical Help from Non-Physician	−1.392**	−.287	.209
$X^2=3.363$***; Pseudo $R^2=.248$ (n=71)			
Cough			
Prefer Drive	−1.680***	−.465	.186
$X^2=8.082$***; Pseudo $R^2=.193$ (=54)			

Notes: ***significant at the .001 level
**significant at the .01 level
*significant at the .05 level

was likely to be a multi-specialty facility. This is in contrast to central Texas where respondents indicated that they had only traveled an average of 3.4 miles to see a specialist and lived approximately 14 miles from the nearest hospital.

Among nonusers, the variables associated with the preference for using telemedicine rather than relying solely on the local physician are more varied (see Table 3). This is not a choice that involves differences in travel but a choice between the local provider alone and the provider with the addition of consultation through telemedicine. For the condition of pregnancy, a preference for the addition of telemedical consultation along with the care of the local provider is associated with less recent contact with a physician, and recognition of the convenience telemedicine offers. For the less serious conditions of cough and rash, the preference for the

Table 3. Logistic Regression of Use of Telemedicine versus Stay for Local Doctor Alone for Various Conditions (Nonuser Patient Sample)

	Use Telemedicine versus Stay for Local Doctor Alone		
Pregnancy	*Parameter Estimate*	*Standardized Coefficient*	*Odds Ratio*
West Texas X Nonphysician Help	-2.718^{***}	$-.551$	.066
Convenience	1.310^{**}	.345	3.708
Male	-1.466^{***}	$-.370$	.231
White	-1.219^{**}	$-.338$	.296
$X^2=23.161^{***}$; Pseudo $R^2=.334$ (n=60)			
Difficult Delivery			
Retired	2.408^{**}	.642	11.111
$X^2=6.337^{**}$; Pseudo $R^2=.203$ (n=45)			
Cancer			
West Texas X Child Present	-1.557^{***}	$-.356$	.211
West Texas X Camera Present	$-.771^{*}$	$-.202$	.460
Time Lived in Community	$-.028^{***}$	$-.331$	.972
$X^2=29.754^{***}$; Pseudo $R^2=.283$ (n=104)			
Earache			
West Texas X Child Present	-3.123^{***}	$-.678$	.044
West Texas X Unemployed	2.691^{**}	.378	14.744
$X^2=20.158^{***}$; Pseudo $R^2=.213$ (n=87)			
Earache			
West Texas X Child Present	-2.303^{***}	$-.498$	.100
No Transportation	$-.977^{**}$	$-.215$	.377
Risk Avoidance	$.138^{**}$	.260	1.149
$X^2=21.258^{***}$; Pseudo $R^2=.222$ (n=117)			
Rash			
West Texas X Prefer Camera	-2.058^{***}	$-.532$	.128
Prefer Personal Contact	-1.780^{***}	$-.415$	.169
Risk Avoidance	$.141^{**}$	.266	1.151
Retired	$.709^{*}$	.178	2.032
$X^2=29.165^{***}$; Pseudo $R^2=.252$ (n=173)			
Cough			
West Texas X Risk Avoidance	-2.319^{***}	$-.640$	.098
Risk Avoidance	1.787^{***}	.396	5.971
Employed	1.062^{**}	.294	2.892
Have Doctor	1.057^{**}	.257	2.887
Seen Specialist	1.149^{**}	.311	3.156
$X^2=31.369^{***}$; Pseudo $R^2=.273$ (n=169)			

Notes: ***significant at the .001 level
           **significant at the .01 level
           *significant at the .05 level

addition of telemedical consultation is associated with the inability to travel (e.g., can't drive), less proximity to the hospital, and less willingness to take risks. Our interpretation of this finding is that these are people who attend more closely to any medical condition and might choose to see a specialist if it were more convenient to do so. For two conditions, other variables were significant: non-white status, female gender (for pregnancy), children in the household, and longer residence in the community (for cancer) are associated with a preference for the physician alone. Non-White status and longer residence in the community are also associated in our sample with lower income, a variable that is not highly associated with adoption generally or with the resources necessary to overcome the distances to specialists.

Table 4. Logistic Regression of Telemedicine versus Travel to See a Specialist for Various Conditions (User Patient Sample)

	Telemedicine versus Travel to See Specialist		
Pregnancy	*Parameter Estimate*	*Standardized Coefficient*	*Odds Ratio*
No Transportation	−1.609**	−.430	.200
X^2=5.333**; Pseudo R^2=.210 (n=80)			
Difficult Delivery			
Employed	1.811**	.106	1.112
Time Lived in County	.106***	.965	1.112
Male	2.909*	.804	18.340
X^2=11.485**; Pseudo R^2=.541 (n=29)			
Cancer			
Time Lived in Community	−.037*	−.404	.963
Income	−.332*	−.332	.718
Male	1.834***	.472	6.260
X^2=9.914**; Pseudo R^2=.232 (n=55)			
Earache			
Prefer Camera	1.639**	.371	5.148
Male	1.343**	.422	3.830
X^2=8.404***; Pseudo R^2.205 (n=55)			
Rash			
No Transportation	1.887**	.326	6.589
Employed	2.550***	.591	12.806
Hispanic	1.908**	.337	6.742
X^2=12.637***; Pseudo R^2=.396 (n=45)			
Cough			
Heard of Telemedicine	1.822**	.504	6.185
Distance to Next Nearest Hospital	−.007	−.303	.993
Recent Resident	−2.659***	−.514	.070
X^2=17.689***; Pseudo R^2=.345 (n=50)			

Notes: *** significant at the .001 level
** significant at the .01 level
* significant at the .05 level

Users

Respondents who had some experience with telemedicine presented somewhat different patterns of choices and variables associated with them (see Tables 4 and 5). For example, responses to the question of whether they would travel to see a specialist or stay in the local community and access a specialist via telemedicine took the following pattern: persons who would use telemedicine for a difficult delivery were familiar with telemedicine and were better educated (see Table 4). For cancer, those who would use telemedicine were risk avoiders and longer-term residents. Those who would use it in the case of a cough were recent residents and lived farther away from the nearest source of specialists.

Table 5. Logistic Regression of Use of Telemedicine versus See Local Doctor Alone for Various Conditions (Use Patient Sample

	Use Telemedicine versus See Local Doctor Alone		
Pregnancy	*Parameter Estimate*	*Standardized Coefficient*	*Odds Ratio*
Risk Avoidance	−.302*	−.397	.739
Time Since Saw Physician	.928*	.519	2.530
Male	1.725*	.455	5.610
$X^2=7.563$**; Pseudo $R^2=.275$ (n=30)			
Difficulty Delivery			
Time lived in Community	.067***	.645	1.070
Male	1.064*	.437	5.122
Age	−.077	−.682	.962
$X^2=12.190$***; Pseudo $R^2=.416$ (n=29)			
Cancer			
Distance Traveled to Specialist	.009***	.935	1.010
$X^2=15.853$**; Pseudo $R^2=.561$ (=74)			
Earache			
No Transportation	2.738**	.690	15.455
Child Present	1.726**	.430	5.615
Recent Resident	−2.508***	−.528	.081
$X^2=12.378$***; Pseudo $R^2=.371$ (n=52)			
Rash			
High School Education or Greater	.947*	.262	2.578
Risk Avoidance	1.182**	.327	3.263
$X^2=7.393$*; Pseudo $R^2=.149$ (n=65)			
Cough			
Heard of Telemedicine	1.098*	.304	2.999
Prefer Camera	1.334*	.307	3.801
Risk Avoidance	.321***	.577	1.379
Recent Resident	−1.856***	−.374	.156
$X^2=23.806$***; Pseudo $R^2=.414$ m (n=65)			

Notes: ***significant at the .001 level
**significant at the .01 level
*significant at the .05 level

The effects of distance and resources (income) are also evident in these data. Some of the variables associated with a preference for traveling to see a specialist are those representing propensity factors (risk avoidance) and enabling factors (recent residents are generally higher in income). It is interesting to note that in the case of cancer, a preference for traveling to see a specialist is associated with longer residence in the community—an indicator of lower income in this sample. However, the risk avoidance factor may be the overriding characteristic. For those who prefer to stay and use telemedicine to access a specialist, male gender, lower income (longer residence, Hispanic ethnicity), inability to travel, and familiarity with telemedicine are all linked to this choice.

When the choice involves using telemedicine vs. relying on the local doctor alone, males were more likely to select telemedicine in the case of pregnancy, difficult delivery, and cancer (see Table 5). Risk avoiders were also more likely to select telemedicine under the conditions of rash and cough. More recent residents were likely to select telemedicine in the case of cough, but to select their local physician if the condition involved a difficult delivery.

SUMMARY AND CONCLUSIONS

Prior experience with telemedicine tended to make patients more willing to choose it under most of the hypothetical health conditions offered. In the case of cancer, however, exposure to telemedicine led to a greater propensity to select travel to see a specialist for this condition. Although experience with telemedicine results in general acceptance of it, conditions such as cancer are viewed as too grave to deal with in distance specialty consultations. Interviews with health care providers indicate they, too, perceive that there are circumstances under which a face-to-face contact is preferable to a teleconsultation (McIntosh, Alston, Mack, Sykes, and Segura 1999).

Among those who had never experienced telemedicine, predisposing factors such as having children were important in several of the models, but enabling factors had more influence on the decision to adopt telemedicine. Predisposition had the biggest impact on more serious conditions such as pregnancy or cancer. Enabling factors were more likely to play roles in less serious conditions such as rash and cough. These factors involved transportation and the distance required to travel for specialty care and were more prone to impact western than central Texas patients. Among those patients who had never experienced telemedicine, predisposing factors played an even greater role in the models for cancer and earache in a child when the choice was between telemedicine and travel to see a specialist. The same, however, could be said for the choice between telemedicine and consulting with the local physician in many of these models. Males, persons who had lived in the community for shorter periods of time, risk avoiders, and Hispanics were more likely to select the telemedicine option. In general only the conditions of a rash or cough were affected by enabling factors such as knowledge of the telemedicine system, distance from specialty care, employment, or inability to drive. Those with greater education, who were employed, who had heard of telemedicine, who were unable to drive, and who lived farther away from specialty care were more likely to select telemedicine for one or more of the conditions compared to either driving to see a specialist or remaining to consult with their local doctor.

Persons who had used telemedicine before differed in important ways from nonusers. Users were less well educated and had fewer resources. They tended to be more geographically isolated in terms of the distance they lived from health care resources. These individuals were also less inclined to be risk takers when it came to health. Such characteristics may be more of a function of residence, as the majority of users came from western Texas than from central Texas. Western Texans were more likely to choose telemedicine over driving to see specialists. Interviews with the primary care personnel in western Texas indicated that no patient had ever turned down the offer to use telemedicine (McIntosh et al. 1999). However, it was the primary care provider who chose which patients ought to receive a teleconsultation. This leads us to suspect that patients may have been chosen by physicians for telemedicine on the basis of known characteristics: the physician knew that the patient lacked resources but was inclined to want to take action when ill or at risk of being ill.

ACKNOWLEDGMENTS

This study was supported by grant number R01 HS08247 from the Agency for Health Care Policy and Research (AHCPR). We would also like to thank Drs. E. Jay Wheeler, M.D., Ted Hartman, M.D., William McCaughan, Ph.D., and Jane Preston, M.D. for their comments on an earlier version of this paper.

REFERENCES

Aday, L.A. and W.C. Awe. 1997. "Health Service Utilization Models." Pp. 153-172 in *Handbook of Health Behavior Research. Volume I. Personal and Social Determinants*, edited by David S. Gochman. New York, NY: Plenum.

Bluestone, Harold and Stephen G. Daberkow. 1985. "Employment Growth in Nonmetropolitan America: Past Trends and Prospects to 1990." *Rural Development Perspectives* 1: 1-42.

Brown, L. 1981. *Innovation Diffusion: A New Perspective*. New York, NY: Methuen.

Fenwick, A.M. and J.S. Schwartz. 1994. "Physicians' Decisions Regarding the Acquisition of Technology." Pp. 71-84 in *Medical Innovation at the Crossroads. Volume IV. Adopting New Technology*, edited by A.C. Geijins and H.V. Dawkins. Washington, D.C.: National Academy Press.

Geertsen, R. 1997. "Social Attachments, Group Structures, and Health Behavior." Pp. 267-288 in *Handbook of Health Behavior Research. Volume I. Personal and Social Determinants*, edited by David S. Gochman. New York, NY: Plenum.

Gochman, D.S. 1997. "Institutional and Cultural Determinants." Pp. 329-336 in *Handbook of Health Behavior Research. Volume I. Personal and Social Determinants*, edited by David S. Gochman. New York, NY: Plenum.

Grigsby, B. and A. Allen. 1997. "Annual Telemedicine Program Review. Part II: United States." *Telemedicine Today* 5(Aug.): 30-42.

Hassinger, E. 1959. "Stages in the Adoption Process." *Rural Sociology* 24: 52-53.

McIntosh, W.A., L.T. Alston, W.R. Mack, D. Sykes, and C. Segura. 1999. Telecommunication Adoption and Use for Rural Health Care. Final Report to the Agency for Health Care Policy and Research. College Station: Center for Public Leadership Studies.

Mintzer, C.L., C.J. Wasem, and D.S. Puskin. 1997. "Program Activity in the Second Year of the Rural Telemedicine Grant Program. Part I." *Telemedicine Today*, 5 (Oct.): 35-39.

Ricketts, T.C. and L.A. Savitz. 1994. "Access to Health Services." Pp. 91-112 in *Geographic Methods for Health Service Research: A Focus on the Rural-Urban Continuum*, edited by Thomas C. Ricketts, Lucy A. Savitz, Wilber M. Gesler, and Diana N. Osborne. Lanham, SC: University Press of America.

Rogers, E.M. 1995. *Diffusion of Innovations*. 4th ed. New York: Free Press.

Wolinsky, F.D. 1988. *The Sociology of Health: Principles, Practitioners, and Issues*. 2nd ed. Belmont, CA: Wadsworth.

INTEGRATIVE MEDICINE:
ISSUES TO CONSIDER IN THIS
EMERGING FORM OF HEALTH CARE

Melinda Goldner

ABSTRACT

A new medical model has emerged due to the public's increasing awareness and acceptance of alternative medicine. As a result, alternative practitioners have joined with physicians in a variety of professional settings to explore ways to integrate Western medical techniques with alternative medical techniques and ideology. For example, clients with chronic lower back pain may receive treatments in an integrative clinic from the physician, chiropractor, and massage therapist. Yet, they are also encouraged to make changes in their daily routines at work and home to lessen the stress on their back. Thus, practitioners use both Western and alternative techniques in accordance with a key component of alternative ideology: the belief that individuals must take responsibility for their health.

Political and cultural changes have made integrative medicine possible, yet there are some key issues that activists need to resolve as they develop this new model of medicine. Many alternative practitioners are interested in working with physicians, because it brings legitimacy to their work. Yet, it is important to understand why

Research in the Sociology of Health Care, Volume 17, pages 215-236.
ISBN: 0-7623-0644-0

some physicians are now interested in working with alternative practitioners. Political changes, such as the rise of managed care, have eroded physicians' authority. Consequently, some physicians are searching for new ways to practice medicine without these structural constraints. Other physicians are drawn to the connection that alternative ideology makes between spirituality and medical practice, reflecting a new cultural emphasis on spirituality. Finally, physicians and alternative practitioners need to develop a team approach where all practitioners have equal power and maintain the ideological integrity behind their techniques. These elements are critical for integrative medicine to be successful and effective.

INTRODUCTION

Activists within the alternative health care movement are beginning to integrate Western and alternative medicine rather than remaining independent from Western medicine. They have opened integrative clinics that encompass alternative practitioners, such as acupuncturists, chiropractors, and massage therapists, as well as nurses and physicians. These practitioners use a combination of Western and alternative treatment modalities; however, they follow the principles of alternative medicine, including the belief that individuals are responsible for their health (Aakster 1986, p. 269). This chapter describes the political and cultural changes that have made integrative medicine possible and it explores some of the issues that activists need to resolve as they develop this new model of medicine. Since many believe that integrative medicine is the "medicine of the future," we need to examine this trend to understand changing and emerging patterns of health care delivery (Dr. Woodson Merrell as quoted in *Cosmopolitan*; Levine 1997, p. 313).

In order to understand integrative medicine, we must first define alternative medicine. Prior theorists have conceptualized alternative medicine as a set of techniques or beliefs, or as a social movement (Lowenberg 1989). Alternative medicine includes well-known techniques such as acupuncture, chiropractic, bodywork, and Yoga, as well as lesser known techniques such as Reiki, which is a form of hands-on energy transmission that is used to balance the life force energy throughout the body. Core beliefs include defining health as well-being rather than the absence of disease, stressing individual responsibility for health, advocating health education, controlling social and environmental determinants of health, and using "natural" therapeutic techniques (Kopelman and Moskop 1981). Defining health as well-being, for example, requires that alternative practitioners go beyond patients' physical symptoms or disease to address their mental, emotional, spiritual, and social well-being (Alster 1989, p. 64; Tubesing 1979, p. 101). Alternative practitioners are attentive to the connection between the mind and the body, or how stress can exacerbate disease. It is also important to view alternative medicine as a social movement. Practitioners and clients identify with

Table 1. Partial Comparison of Western and
Alternative Medicine

WESTERN MEDICINE	ALTERNATIVE MEDICINE
The focus is on disease and limitations.	The focus is on health and self-control.
Focus on how all patients with an illness or symptom are similar.	Focus on individual uniqueness despite similar illnesses or symptoms.
Disease is caused by an invasion of the body by outside agents such as germs.	Disease is caused by a distortion of the self-healing and self-regulating systems in the body.
Diagnose through observation and medical tests.	Diagnose through subjective evaluations of clients.
Diagnosis identifies the causative agent.	Diagnosis identifies life factors.
Control or eliminate symptoms in order to manage disease.	Restore the homeostatic balance in the body in order to eliminate disease.
Drugs and surgery are used to suppress or eliminate symptoms.	Drugs and surgery are avoided since they further disrupt homeostasis.

Note: Adapted from Coulter in Sobel 1979, p. 294; Lowenberg 1989; p. 34; Mattson 1982; pp. 49, 126-127; and Weiss 1984; pp. 141-142

an alternative health care movement and define their participation as activism, and it is this collective effort that has led to recent changes in both Western and alternative health care organizations (Goldner 1999; Lowenberg 1989, p. 73; Schneirov and Geczik 1996, pp. 640-641). Table 1 provides a comparison between the alternative and Western medical models.

The dramatic rise in both the interest in and use of alternative medicine is evidenced by the increase in news stories and Web sites devoted to alternative medicine, as well as the amount of homeopathic remedies and supplements in the local supermarket[1]. The most widely cited statistics come from two studies led by Eisenberg in 1993 and 1998. Their original study (Eisenberg, Kessler, Foster, Norlock, Calkins, and Delbanco 1993) was groundbreaking because they showed that more individuals were using alternative medicine than previously thought (34%). Their follow-up study (Eisenberg, Davis, Ettner, Appel, Wilkey, Van Rompay, and Kessler 1998) continued to show extensive use; 42% had used an "unconventional therapy" in the past year (p. 1569). In the most recent study, respondents were most likely to have tried relaxation techniques, herbal medicine, massage, and chiropractic (16%, 12%, 11%, and 11% respectively); whereas, they were least likely to have tried hypnosis, biofeedback, and acupuncture (1% each)(p. 1572). Use was still more common among college educated, higher income 35- to 49-year-olds, especially women and those of all racial groups besides African Americans, just as it was in the original study (p. 1571). Patients were still more likely to seek treatment for chronic, rather than acute and/or life-threatening medical problems. The majority still did not tell their physician about this use (61.5%) and paid the entire cost of treatment out of pocket (58%)(p. 1573). Since the *New England Journal of Medicine* and the *Journal of the American Medical Association* published these studies, it is likely that medical doctors as well as alternative practitioners know about the findings. In fact, one respondent in my study found that physicians "quote Eisenberg a lot. That [study] was in the *New England Journal of Medicine*, and it's their premiere journal."

Proponents believe that integrative medicine combines what is best about Western medicine, the techniques, with what is valuable about alternative medicine, the techniques and, more importantly, the ideology. By retaining what is useful in Western medicine, proponents argue that integrative medicine improves upon alternative medicine in both ideology and practice.

The first reason they believe that Western and alternative medicine should be combined is that to them it seems to be more beneficial than choosing either one exclusively. This combination is what allows practitioners to adhere to the belief that they must address the patient's physical, mental, emotional, spiritual, and social well-being. Given this belief, practitioners need to use multiple therapies since they address so many facets of the person and do not believe that illness has a single etiology (Alster 1989, p. 61). The director of an integrative clinic, a respondent in my study, explains that these practitioners recognized "their limitations...I think a lot of these groups are forming to combine resources and cover more of the bases. We recognized that people need the whole spectrum [of resources] to get better and stay that way." Proponents of alternative medicine often criticize the manner in which physicians use Western techniques, not always the techniques themselves. This is illustrated by the following comments from two physicians observed in this study. One said, "don't put down Western medicine. It's useful. It's why we are living longer." Another said we would be "foolish to throw out Western medicine." Western medicine is useful for diagnosing and monitoring conditions as well as treating acute crises such as broken bones or infections. Western medicine's best "tools" are drugs and surgery, yet there are limits to relying on these solely. Alternative medicine is better for chronic illnesses or those illnesses where lifestyle plays a role, such as heart disease. Thus, both alternative and Western techniques are necessary to heal people.

There is a second reason for combining Western and alternative medicine in addition to the fact that the patient benefits more from the combination than from either by itself: practitioners and activists know that clients *want* both Western and alternative medicine. Most individuals use alternative techniques in conjunction with, rather than instead of, Western medicine (Eisenberg and colleagues 1998, p. 1573). Another respondent in this study says she wishes "Western and alternative medicine could mix more. I don't see it as black and white, or good and bad, just as complimentary. Both sides are weary of the other, but I think there is a place for both."

In this article I examine why this new medical model has appeared, and identify some potential problems that must be resolved in order for it to be successful. Specifically, I examine why alternative practitioners, consumers, and most importantly some physicians have begun to advocate this model. Then, using data from a larger study on alternative and integrative medicine in California, I highlight how physicians and alternative practitioners need to develop a team approach where each practitioner has equal power and maintains the ideological integrity behind their techniques.

METHODOLOGY

I conducted an ethnographic study of the alternative health care movement in the San Francisco, California Bay area between 1996 and 1998. Numerous scholars and advocates define this location as the center of activity for the alternative health care movement (Berliner and Salmon 1979). I base this qualitative ethnographic study on interviews, secondary analysis of movement documents, and participant observation (Van Maanen 1982).

Initially, forty alternative practitioners and clients were interviewed. These interviews served as the primary source of data. I used previous literature and the local telephone directories to compile a list of alternative clinics in the San Francisco Bay area. I then called local clinics and posted fliers in various locations to find respondents. Since this technique solicited practitioners almost exclusively, I used snowball sampling to gain access to clients. Of the 40 respondents, 30 were alternative practitioners (75%). Though at the time of the study the remaining 10 respondents were solely clients of alternative techniques (25%), two of these individuals were in training programs to become alternative practitioners but had not finished.

The sample includes only those individuals who participated in interviews. Overall, respondents were overwhelmingly female (73%) and Caucasian (97%), and ranged in age from 35 to 63 (mean age=47)[2]. All respondents had taken some college courses, and 71% finished some graduate work or received graduate degrees. Religious or spiritual identification varied greatly, though 26% said they had no affiliation whatsoever. Of the respondents, 42% are currently married, though an additional 33% were previously married. Finally, respondents did not report their incomes accurately enough to ascertain a reliable range or mean. Only 50% of the sample gave enough detail to provide the following information. Among those 20 respondents, annual earnings were anywhere from $18,000 to $80,000 in the year prior to our interview (mean income was $41,980). Of those not included in the range or mean, however, one respondent said he made "six figures," one reported an income of only $3000, and another earned $8568 on disability.

Next, following the interview process, I conducted secondary analysis of movement and clinical literature to increase my knowledge of the local and national movement (Jorgensen 1989, p. 22). These include newspaper and magazine articles, activist newsletters, event announcements, position papers, and clinic handouts describing services. This was not a random sample, nor were these documents representative of the movement as a whole. Rather, they helped me understand what was happening on a national level and locally within the San Francisco, California Bay area. I examined articles from sources as diverse as *Alternative Medicine Digest* and *TIME*. Many of these documents, especially mainstream sources, allowed me to see how prevalent alternative medicine was. I did not limit these sources to published material because I was also interested in literature that would be given to individuals who were already patients of alternative clinics or advocates of alternative medicine. Though not a primary source of data, I used these data to supplement information from the interviews and observations.

Finally, I observed three alternative clinics, one activist organization, and one professional organization. The clinics are service-oriented since clients consult practitioners in these settings for advice and treatment. The primary task of practitioners is to deliver health care, even though they also educate and empower clients. The activist organization focuses on educating and lobbying rather than providing services. This particular organization has a professionalized national organization with decentralized local organizations. The professional organization is similar to any group of professionals that form an organization to educate members and advocate on their behalf. While lawyers have the American Bar Association and physicians have the American Medical Association, this professional organization contains both alternative and Western practitioners. They educate each other and advocate integrative medicine. Even though clients periodically attend case conferences on specific illnesses, the practitioners do not deliver services to these clients within the confines of this organization so it is not service-oriented. For the purposes of this paper, I describe only the integrative clinic and professional organization in more detail.

Integrative Clinic

The integrative clinic combines Western and alternative medicine based on the belief that both types of medicine serve a purpose and have a place in patient care. Its goal is to provide "comprehensive health care that works better than any one approach by itself" (clinic flyer). Though practitioners pay rent, provide their own malpractice insurance, and retain the entire amount they collect from clients, practitioners advertise under one collective name and act as a team. For example, practitioners need approval from the co-directors for any advertisements since it has the clinic's collective name on it.

I observed an integrative clinic that provides clients a range of techniques. Their services include acupuncture, homeopathy, herbology, internal and family medicine, women's health, nonforce chiropractic, bodywork, Yoga, pain management, psychotherapy, lifestyle guidance, nutritional counseling, biofeedback, and movement re-education. Their staff offer many additional services: (1) preventive, internal, and sports medicine; (2) menopause, gynecological conditions, and infertility treatment; (3) neurolinguistic programming; and (4) clinical hypnotherapy. Since there has been some staff turnover, this range of services varies considerably over time.

Given the range of techniques, the clinic has different funding sources, which affect their work. The clients' form of payment depends on the practitioner they see. Third-party payments include worker's compensation and auto insurance but consist primarily of private health insurance. One practitioner, who utilizes acupuncture and homeopathy, receives less than half of his payment from third parties, whereas the physician and nurse receive more than 90% in this way. The chiropractor would be somewhere in between these two extremes. Overall, most clients "pay out of pocket," according to the director. They have actually found that when clients do not pay for their services directly, they are less likely to take responsibility for their healing. As mentioned before, this is a fundamental component of alternative ideology.

Briefly, their clientele have a broader range of conditions, ideologies, and willingness to try different techniques. Some clients are drawn either by Western or alternative medicine, and use these techniques exclusively. For example, some clients only request Western techniques such as a yearly gynecological exam. Others have a condition and are open to various techniques. For example, someone with acute back pain may be willing to try anything from pain management through Western medicines to acupuncture or movement re-education. It is an advantage for this latter group of clients to obtain all of their medical care in the same location and know that their practitioners consult each other. This broad range of clientele enables the integrative clinic theoretically to have more clients than a clinic that focuses exclusively on Western or alternative medicine.

Integrative Professional Organization

I observed a professional organization devoted to integrative care. Several physicians started the group, and one acts as the director. Within a year and a half it had expanded to nearly 200 members, consisting of physicians, alternative practitioners and a small number of lay people (e.g., health care entrepreneurs). In one newsletter, they said the group consisted of over 60 physicians as well as various chiropractors, acupuncturists, naturopaths, and bodyworkers.

Most of the monthly meetings revolve around education and awareness. Speakers discuss a variety of topics from nutrition to how to generate income from preventive health care. They also educate members through a monthly newsletter.

For example, they include highlights from the speakers and advertisements in these issues. They spent the first 18 months trying to understand and educate each other. This process was extremely difficult since practitioners typically have a different ideology, set of guiding assumptions, and language underlying their work. This dialogue culminated in a two-day professional symposium on integrative medicine during June 1997.

I started observing this group while they were in the final stages of planning this symposium. Their goals were to establish communication and understanding between diverse professionals (Western and alternative), explore the benefits to patients and providers of integrated care, learn how to view the whole patient, and make the shift from disease care to health care. They organized speakers around the four areas of the organization that include the spiritual, physical, biochemical, and energetic. Approximately 125 participants attended the symposium, though it is unclear how many were not members of the group directly.

In addition to observing and interviewing members, I videotaped the symposium (Jorgensen 1989, p. 22). An anthropologist studying this organization also gave me videotapes of 8 monthly meetings, 6 leading up to the symposium and 2 following it. Consequently, I have information on this group's activities from January through September 1997. Van Maanen (1982, p. 103) includes filming as an ethnographic strategy.

EXPLAINING THE INCREASED INTEREST IN INTEGRATIVE MEDICINE

Integrative medicine is now feasible due to this increased interest in alternative medicine, as well as new political and cultural opportunities. Politically, there is an "apparent loss in a single decade of the authority that medical professionals held for much of the twentieth century" (Mick 1990, p. xiii). Overall, the medical environment is increasingly complex (Gray 1986, p. 172). There is no longer a single, dominant group or figure in the medical industry because power has moved from physicians to the medical schools, hospitals, health insurance companies, health care chains, and financing and regulatory agencies (Starr 1982, pp. 8, 421). Now the medical sector resembles other economic sectors that are subject to market pressures. This contrasts with previous periods when physicians exemplified unparalleled power and authority (Scott 1993, p. 281). Because physicians are losing their authority due to these political changes within Western medicine, they are willing to explore integrative medicine.

Medicine has been transformed into a health care industry where profit, efficiency, financial managers, and consumers take center stage. We have seen changes in ownership and control of health related industries, specifically toward more concentration, now that investors believe health care is a profitable enterprise (Scott 1993, p. 275; Starr 1982, p 429). More private companies own and control health care organizations. For example, hospitals are now part of corporate organizations (Scott 1993, pp. 275-279). In 1983, U.S. Health Care Systems became the first health maintenance organization (HMO) owned by investors. By the mid-eighties, one-third of all HMO's were for-profit (Bloom 1987, pp. 52-53). Today, 12 large corporations (e.g., Kaiser, CIGNA, Blue Cross/Blue Shield) account for most HMO memberships (Bloom 1987, p. 91). The locus of control moves away from freestanding, local structures and individual physicians to non-local, multilayered authority structures (Mick 1990, p. 2; Starr 1982, p. 429).

Specifically, physicians are losing authority because financial managers are gaining influence within health care settings (Scott 1993, p. 279). Since the environment is increasingly complex and competitive, business professionals often manage health care facilities (Gray 1986, p. 172). We assume corporations are better able to handle the complexity of the environment where health care facilities now operate (Scott and Backman 1990, p. 34). Managers assume more power and control since the emphasis is increasingly on profits. Earlier theorists thought health care was immune to market pressures (Scott 1993, p. 280). Now we are seeing changes within medicine that reflect the opposite. At a basic level, the language has changed from discussing a medical care system to a health care *industry* (Mick 1990, pp. 6-8; Scott 1993, p. 280). One physician in this study says that medicine has become an "explicit business." This new language reflects how profit, efficiency, rationality, and consumers have taken center stage. To maximize profits, HMO's and other insurance companies are changing their reimbursement policies (Mick 1990, p. 4). Many physicians find themselves trying to make their diagnosis fit the company's reimbursement policies, or having an administrator question their professional judgment (Weitz 1996, p. 237). Physicians, then, become more dependent on financial managers due to these fiscal controls (Scott 1993, p. 279-280). These changes increase managers' power and decrease physicians' power within medical institutions (Gray 1986, p. 172).

Even though these changes entail a gradual erosion of autonomy, physicians are more compliant now and less able to resist these changes. Most importantly, there is an increased supply of physicians so individual doctors lose their bargaining power (Gray 1986, p. 174; Mick 1990, p. 5; Scott 1993, pp. 279-280). Physicians are also more compliant due to the hospital's ability to hire and fire physicians, increased loans from medical education that make physicians look for job security in hospitals and HMOs as opposed to private practice, increased costs due to fear of malpractice and the consequent attempts to prevent it, fewer hospitals in which to practice, and more alternative delivery systems that make physicians more attentive to patients' needs since they now have competition (Gray 1986, p. 172, 174; Mick 1990, pp. 5-6).

Given the resulting dissatisfaction from their loss of authority, more physicians are turning to integrative medicine for a solution. Just as alternative medicine becomes attractive to consumers, some physicians may begin to see alternative ideology as a mechanism for improving upon weaknesses in Western medicine. It is unlikely that physicians would abandon their knowledge to learn a new alternative technique. These physicians also realize that Western medicine is useful in some cases. Consequently, some physicians may (1) simply refer their patients to alternative practitioners while others may (2) change the way they practice Western medicine.

In the first option, physicians might recognize their limitations and refer patients to alternative practitioners. A dance therapist in this study has seen a change in physicians' reactions. She says that physicians have learned that "nothing works universally." One client says her physician feels frustrated when she cannot help her clients medically.

> *She says the hardest thing as a doctor is to have this revolving door of a number of patients [where] all [she] can do is give painkillers or some kind of maintenance drug. For example, doctors can do very little for people with backaches. They can help the patient get over this bout of back pain, [but] it's back in six months; [whereas,] many of the alternatives help people to get rid of that on a permanent basis.*

Physicians might begin to see their techniques as only one method of healing patients, and refer patients to alternative practitioners. Patients benefit from multiple healing modalities, but do not necessarily receive integrative care since the practitioners may not consult with one another. In addition, the patient may only learn about alternative ideology through the alternative practitioners, not the physician.

Option two, and the one more significant to the model described in this paper, is that physicians driven by alternative ideology would change the way they practice medicine. Though these physicians are in the minority of physicians as a whole, they are a growing force within medicine and gain a lot of recognition for their interest. Dr. Christiane Northrup (1994), a leading expert on women's health, says she kept telling herself through medical school to learn the tools of the trade even though she knew physicians practiced them very badly. Physicians may open themselves to alternative ideology, such as believing that individuals are responsible for their own health, in addition to advocating alternative techniques (Konner 1993, p. 15; Tubesing 1979, p. 21). These physicians would change their practice of medicine to include education and empowerment as we have seen alternative practitioners do. Some physicians may incorporate information on the connection between the mind and the body into their work. For example, Dr. Herbert Benson says "people are dissatisfied with routine medicine that's strictly drugs and surgery, and here, now, is a way to incorporate value systems that people cherish [into medical practice] without sacrificing science" (as quoted in Baker 1997, p. 20).

In addition to the political changes within Western medicine, cultural shifts make integrative medicine more desirable. This includes an increased emphasis on spirituality (Dossey 1989, 1996; Gawain and King 1986; Jampolsky 1970;

Norwood 1994; Peck 1997; Williamson 1994; Zukav 1989). Williamson (1994) suggests "a mass movement is afoot in the world today, spiritual in nature and radical in its implications" (xvii). This movement is fueled, in large part, by the baby boomers (Roof 1993). Baby boomers, due to their sheer size, have influenced cultural trends since their involvement with the social movements of the 1960s and 1970s. According to some authors, they are now reaching mid-life and are trying to reconcile aging bodies, impending mortality, and economic insecurity with their sixties' beliefs and independence. These cultural shifts make integrative medicine more desirable, because integrative practitioners examine the relationship between spirituality, health, illness, and well-being.

Some physicians are drawn to the connection that alternative ideology makes between spirituality and medical practice. This study examines an organization of medical professionals exploring the connections between spirituality and professional practice, or integrating one's "inner" and "outer" practice. Their emphasis that "who they are" is more important than "what they do" might be comforting to some physicians, given the frustrations physicians now have with their work due to structural changes such as managed care (organizational newsletter). One physician wrote in this organization's newsletter that he initially explored alternative medicine for the additional healing tools but soon came to understand that it was more than the techniques that could benefit his patients. Understanding how his spiritual practice could benefit his medical practice helped him connect with, and heal, his patients more effectively.

Some physicians are drawn to the spiritual aspects of integrative medicine, especially since some believe that Western medicine has forgotten how to heal people. Activists argue that Western medicine focuses on *curing* at the expense of *healing*. Dr. Dean Ornish says that "to me 'curing' means only getting back to the way we were before we became diseased. 'Healing' is when we use our pain or illness as a catalyst to begin transforming our lives" (as quoted in Easthope 1993, p. 292). Since chronic diseases are increasingly prevalent and since many diseases result in death, the movement's emphasis on spiritual healing has appeal. Activists recognize that people can heal without having a medical practitioner cure the disease. So even though some individuals cannot change the reality of the disease, they can keep their reaction to it within their control. A physician in my study said he likes the "profit potential" of integrative medicine, but knows it is also "the right thing to do" since people need healing.

In theory, physicians who are interested in the ideology of integrative medicine should be more likely to give up their power to work with alternative practitioners as equals. This is because through this ideology they will have learned that they cannot help patients on their own. One physician in this study said he doesn't "have time to learn about" all of these other alternative techniques. If physicians pursue alternative medicine with the ideology intact, then physicians and alternative practitioners could develop integrative models where they have equal power. Unfortunately, this does not happen in practice. An imbalance of power between alternative practitioners and physicians is just one area that activists need to address before an integrative model of medicine will be effective.

ISSUES

Since integrative medicine is so new, it is important for observers and consumers to understand some of the issues that activists must resolve in order to develop an effective medical model. I do not provide a comprehensive list of potential problems due to space limitations; however, the specific issues I raise are representative of broader issues within the movement. The first to be highlighted are potential problems that activists have in developing a team approach with diverse practitioners.

Developing a Team Approach

Practitioners need to learn different techniques and develop a team approach to make an integrative clinic operate. Finding practitioners that can work together as a team has been one of the biggest difficulties practitioners have faced within the integrative clinic. The director says this requires "people letting go of their control. All of us practitioners, especially physicians, are used to being in control of that patient [and having] the final word." Perhaps explaining why the majority of the practitioners in the integrative clinic are women, the director finds that "women find it easier to not have to be the one in charge of calling the shots."

In addition to finding practitioners who can let go of their "ego" and control (Tubesing 1979, pp. 128, 147), the integrative clinic needs people who can

> understand each other and communicate. Each discipline
> has their own vocabulary, [and] their own set of principles.
> How do you get these people to really understand, appreciate,
> and [know] when to use each other? Well, we are tackling
> this by doing it. We are developing a common language.
> We're meeting together, and educating each other. We're
> having case conferences, and experimenting basically. All
> of us are pretty familiar with what results we can get on our own.

Likewise, the professional organization spent the first 18 months trying to communicate. They also experienced difficulties since their backgrounds were so diverse. Yet, these difficulties do not just lie between physicians and alternative practitioners. One member said that the organization helps alternative practitioners become familiar with each other's work since they are not necessarily knowledgeable about other alternative techniques. In other words, acupuncturists may be just as unfamiliar with the work of massage therapists as physicians are with the practice of acupuncture.

Once practitioners have developed a common language they need to learn how to work together on individual cases. The integrative clinic plans to have several practitioners interview the client at once, and then consult with each other to make treatment recommendations. Since the director of the integrative clinic is a homeopath, which entails an exhaustive intake process, he will "function as one of the overview practitioners [who] sees people for the first time." The physician will also play this role since Western medicine is particularly good at diagnosing and ruling out medical conditions. Clients will choose their treatment plan after practitioners present several options and state their preferences. Practitioners would continue to evaluate and monitor a client's health and treatment program through team conferences.

The integrative clinic has not fully implemented this team approach in practice. They have been informally scheduling team conferences, but would like to make this process more formal. One of the practitioners believes that they are further along with educating each other, rather than "working in a true team process." As it stands now, the practitioner who has the initial contact with the client acts as the primary provider. Other practitioners might become involved in the case; however, as a "courtesy" they should first consult with the primary practitioner. This model is closer to managed care than the integrative model activists desire. Practitioners have not worked out details of the team approach because it is so difficult given the problems just described.

Members of the professional organization are finding similar difficulties developing a team approach even though their organization does not act as a clinic. Members of the professional organization are developing a team approach through examining individual cases and seeing how multiple practitioners could help the client. For example, they have plans to look at AIDS and chronic fatigue syndrome. "I think taking cases is a good direction. It really focuses things so we don't just talk about vague generalities." Yet, during some workshops the presenters seemed to talk past each other, rather than have a true dialogue. They would most often raise more questions than answers. One presenter at the symposium said "none of this is easy, but none of this is impossible." Some members have come away from these meetings no longer believing integrative medicine is even possible. Most of the others lie somewhere in the middle. Illustrating this middle ground, a member of the audience at one of the workshops said that there has been a "great deal of hostility on both sides historically." Yet, she can say that the integration of Western and alternative medicine is "plausible" after participating in this group.

Even though team coordination is time consuming, difficult, and goes against much of one's medical training, it can be rewarding for practitioners and beneficial to patients (Tubesing 1979, pp. 128, 178). It entails "very new territory [which] is part of the excitement that draws practitioners to us", says one clinical director. The chiropractor within the integrative clinic likes the "camaraderie. If I'm having a problem with a patient I can ask other practitioners [for] suggestions.

You get a perspective totally different than you would have imagined. I've learned a lot about other modalities." The nurse practitioner agrees that the camaraderie is the biggest advantage to the team approach. "My sense is that we appreciate hearing each other's input, and that we are learning something [with] every kind of medical dialogue. It keeps it exciting." This team approach also benefits patients. Rather than just referring clients to other practitioners, practitioners coordinate "these approaches in a way that will bring quicker and longer lasting results." The director of the integrative clinic explains that "there is a synergy that can happen. It's not just a matter of 'one plus one equals two'. It can sometimes equal a lot more. If you can integrate your efforts, you can get much better results [for the patient]."

Managing Power Differences

Physicians and alternative practitioners need to have equal power within an integrative model of medicine. Yet, physicians and alternative practitioners have different levels of power in society, and this can hinder a true team approach. At the integrative clinic, one practitioner mentioned that the "physician here likes to be the physician and wave her title. So I have to be respectful, but not necessarily back down." It is clear that the physicians had more power in the professional group. On several occasions, members thanked physicians for attending workshops, especially since they were "putting themselves out there" when "they don't have to [do this]." The alternative practitioners seemed so appreciative that physicians were giving them any level of credibility, that they did not seem to mind differences in power[3]. One member sums this up.

> The MD's have the most influence in the group. Unless you do it that way, though, you wouldn't get their participation. They are at the top of the food chain, and you want them involved. That's what's different from holistic medicine in the past. There is a psychic and holistic medicine fair each year. Physicians wouldn't be within one mile of that! If you are a practitioner of Alexander therapy, you can be a part of that one and these other [groups]. There's not much risk. So I think the alternative practitioners are very grateful the physicians are here, and that they are getting something closer to an equal footing. They are very appreciative the physicians are part of the group, and they try to rise to the occasion.

Alternative practitioners are "grateful," because they realize that physicians still have much to lose if they promote alternative medicine. The cultural and political contexts are changing, but opposition within Western medicine remains strong. One respondent has found that physicians are not necessarily skeptical of alternative medicine but "cautious" since they are afraid of "censure and lawsuits." Medical boards can revoke a physician's license for using or recommending alternative techniques in some cases. "So I have found that when

I talk to physicians, they almost feel that they have to be fairly quiet about what they are doing, for fear of censure by their own peers."

Consequently, physicians still have more power than alternative practitioners. Power differences will persist as long as there are different risks involved with joining an integrative practice. Physicians will be able to stay in a position of power within integrative practices since they have more to lose, thus more say in how they develop. We saw that alternative practitioners concede this power since they are grateful that physicians are taking such a risk. Alternative practitioners are coming from a position of powerlessness since they are not taking as much of a professional risk. When you start from a disadvantaged position, you have little to lose.

Differences in power cause some alternative practitioners to use caution when they interact with physicians, especially since physicians are already feeling threatened due to the financial and organizational changes briefly outlined. According to one member of the professional organization, for example, physicians are "threatened by HMOs." Given this, another member of the professional organization said a speaker on alternative medicine was too arrogant and "in your face." There is a "time and place" for that approach, but he believes it is better to persuade physicians through "logic and reason" since it is less threatening.

These power differences between alternative practitioners and physicians at times are connected to financial issues. Many respondents noted that physicians have access to more clients since their services are reimbursed by insurance. Though this is changing, most insurance companies do not cover alternative techniques, or their coverage is very limited. Since alternative techniques are often expensive, this lack of coverage prevents many individuals from using an alternative practitioner's services. As one respondent notes, "right now some call me up and ask if Rolfing will be covered and when I say no, they don't come in [to receive services]." Alternative practitioners, recognizing that this puts them at a disadvantage in relation to physicians, are lobbying insurance companies for coverage. In fact, this is a key strategy of the integrative clinic, which I will describe next. Yet, this only describes power differences between alternative practitioners and physicians on a broader level. It is difficult to tell, however, to what extent these financial issues affect interactions between the two groups on an interpersonal level. Most respondents were so reluctant to admit power differences that it was hard to get them to acknowledge a problem, let alone identify the cause. Future research needs to examine this probable link between reimbursement and power issues between the two groups on a more micro level.

Integrative medicine requires a team approach to health care where practitioners all have equal power in recommending treatment possibilities. For this reason, activists will need to address the existing power differences between physicians and alternative practitioners. Moreover, if they are reluctant to admit these power differences, activists will have an even harder time addressing this critical issue.

Maintaining Ideological Integrity

Activists need to maintain the connection between alternative techniques and the beliefs that guide their practice. However, some alternative practitioners have changed their language to fit into mainstream medicine. The integrative clinic is trying to persuade insurance companies and businesses to send their employees to them for health services, as well as cover both alternative and Western treatments. They tell these companies that they focus on providing "effective, efficient and affordable" care. By emphasizing results and cost-savings, this clinic is trying to work within the existing cultural framework and speak the language of these companies[4]. They do not discuss their ideology, because they do not want employers or members to think that they need to believe to get results. The director said he approaches companies by saying "forget all of the philosophy. We can help your employees stay healthier..." He says

> I'm very careful about talking about this in general. When I speak in public it's more about building bridges with the mainstream, [and] trying to be very, very inoffensive without compromising [the] basic principles. We are not going to talk about the female force, the life force and even spirituality. It's great because it works. So, if you want results, just try it. Now how it gets the results, I think eventually you can't avoid the issue.

Practitioners avoid discussing the actual alternative techniques, such as acupuncture, by saying the clinic simply promises results. When providing more detail on their treatment programs, however, alternative practitioners do need to discuss the different types of techniques they use. So they begin with "conventional medicine" since "it's where most are standing now." It "opens the door to the rest of the system." Possibly contradicting their insistence that the team of practitioners are equal players, one flyer says that the physician makes the initial diagnosis, "prescribes the optimal treatment plan," and "oversees the program." For example, they say that "individual practitioners consult with the physician monthly for team progress reports." Since they are still developing these team conferences, it remains to be seen whether they are simply saying that the physician is in charge to reassure companies and employees or if this is actually the way it will function. One clue we have as to how this will play out is the fact that the one practitioner mentioned how the physician likes to "wave her title." When discussing alternative techniques, they continue to emphasize results. They say traditional Chinese medicine has been "developed over thousand of years, [and] proven through centuries of clinical results"; "herbal medicines are time-tested, and prescribed according to precise guidelines"; and "the role of balanced health is well established" (draft of a flyer).

Even though activists might weaken their discourse, they will not compromise their ideology. Practitioners in this integrative clinic stress that they do not want to treat all of the employees or members within a company since some will not take individual responsibility for their health. Even though they may not discuss their ideology with these companies, they will not compromise their ability to use these beliefs to guide their practice. The director says existing models have made these compromises. For example, he believes that Andrew Weil, one of the leading advocates of integrative medicine, simply lays out a "smorgasbord" of techniques divorced from their ideology. So to maintain this integrity, the clinic needs to remain small and personal so that practitioners can know each other and the client intimately. They are experimenting with a new structure that can accommodate the team meetings they will need to discuss each client separately. They must remain committed to core principles but flexible in how they are manifested since each case will be different. Looking ahead, they realize that the potential exists that they may have to grow faster than they should. Rather than creating a large medical industry, the director envisions small franchises because "it's community that works."

Even when members of the professional organization are educating physicians about alternative medicine, they face the problem of keeping the techniques connected with their beliefs. One of the members said material in their presentations is not new to most alternative practitioners, but to the 20% who are physicians, it is all new information[5]. Some members expressed frustration that physicians in the professional organization seem more interested in learning the techniques than practicing them with alternative ideology intact. To illustrate, physicians often want practitioners to explain alternative techniques in Western terms since they have Western scientific training. One member of the professional organization in particular felt frustrated that symposium organizers wanted speakers to explain their specific alternative technique in a 20- or 30-minute presentation. Rather than explaining a technique in the context of how to use it, this type of presentation educates physicians about alternative techniques, not ideology. Alternative practitioners do not criticize this practice with other members of the professional organization, even though they shared their frustration with me. Many are simply willing to concede some power since they want physicians involved.

Activists are willing to focus on cost-savings and results with insurance companies and physicians, rather than ideology. Since ideology has held this movement together and since it is the key factor that sets their vision of medicine apart from Western medicine, these activists know that they cannot sacrifice having these beliefs guide their practice of medicine[6]. Maintaining this ideology is critical because, without it, physicians may think alternative medicine is only a set of techniques. With the ideology, alternative medicine becomes much more. The director of the integrative clinic explains:

If you get into what I call a truly holistic way of seeing the world, practices come out of that. It will include holistic methods, but it's more how you approach a problem, and asking what does the person need in a holistic way as broadly and deeply as possible. Just to do that evaluation you need a lot of perspectives. Another aspect is you have to have people who get this approach well enough so that they can start to work in it. That's been a big challenge, because it's a new way of thinking. Even if you were trained as an acupuncturist where it's inherently holistic, you can be using it [in a] Westernized [way]. They're still thinking symptoms. You can't just use a holistic tool in the other way. You won't get results. [You need] to really start thinking and acting according to holistic principles. It's a high calling in my opinion. So you get people that have embraced this genuinely in their lives and understand it.

Most importantly, an understanding of the ideology should prohibit physicians from simply co-opting alternative techniques for their practices within Western medicine. Activists teach physicians that multiple practitioners are required to heal clients and hope that physicians will learn there is too much for one practitioner to learn, even a physician. Yet, by moderating their discourse on ideology to gain access, they are, in effect, gambling with the one element they can least afford to lose.

Activists also have to learn how to moderate their discourse on spirituality without compromising their belief that well-being requires spiritual healing. Though the spiritual dimension draws some people to integrative medicine, it may dissuade others, especially physicians, from working with alternative practitioners. To illustrate, not all members of the professional organization are comfortable with the spiritual and communal aspects of the group. The director of the group acknowledged that as a physician, it felt strange to participate in spiritual practices, such as drumming, chanting, and meditating. Another member of the group said he and others often poked fun at the more marginal healers in the group that bring in a spiritual element more explicitly. Another member, an herbalist, said that the spiritual components of integrative medicine may make them lose credibility as a group. Though group members "emphasize whatever one believes" it is obviously difficult for some to appreciate this spiritual diversity, especially since members want to look credible in the physicians' eyes.

Some physicians may be uncomfortable with the spiritual aspects of integrative medicine since Western medicine is a "relatively secular profession" (Peck 1997, p. 202). Dr. Larry Dossey (1984) says that physicians are no less spiritual than others. They just believe religion and science are incompatible or that religion is unnecessary to science (vii). Another argument is that "fewer than two-thirds of doctors say they believe in God. 'It is very important to many of our patients and not that important to lots of doctors'" (David Larson, a research psychiatrist at the National Institute for Healthcare Research, as quoted in Wallis 1996, p. 64). "Western medicine has spent the past 100 years trying to rid itself of remnants of mysticism," or separating medicine from religion (Wallis 1996, p. 59). One respondent, not involved with the professional organization or integrative clinic, said only four physicians came to his talk on "spirituality and health care" even though the "room will be full" when he discusses alternative techniques. Whether physicians believe in the connection between the mind, body, and spirit or not, they may be uncomfortable integrating spirituality into their work. Yet once again, the important issue is that activists need to maintain this component of their ideology; however, they need to moderate discourse about spirituality and recognize that some physicians may be uncomfortable with this element of alternative medicine.

CONCLUSION

How the alternative health care movement handles these issues will influence the level of success of integrative medicine. For an integrative model of medicine to be successful, activists need to develop a team approach where each practitioner has equal power. If alternative practitioners continue to deny that power differences exist since they are simply grateful that physicians are working with them, they will not develop effective models. Activists will also have to decide if they are willing to continue risking the separation of their ideology from alternative techniques by moderating their discourse. They need to realize that they risk co-optation of alternative techniques by Western medicine. Physicians may simply use alternative techniques, such as acupuncture, as they would Western drugs. For true integration, though, "you can't take so called alternative tools and scab them on to the existing system. It won't function, because the whole operative principles are different." Since integrative medicine "takes in a whole world view," you need to blend alternative and Western techniques into a system that keeps alternative ideology intact. "If you're not really applying holistic principles, you're not using these [techniques] to their full potential, and you're not doing your best for the patient." For a true integrative model of medicine, activists need to maintain control of their techniques, make sure not to divorce these techniques from their ideology, and learn how to lessen the power differential between them and physicians. If they can do this, "the integration of alternative and traditional Western medicine [will be] the medicine of the future" (Dr. Woodson Merrell as quoted in *Cosmopolitan*; Levine, 1997, p. 313).

NOTES

1. Web sites include "Ask Dr. Weil," Yahoo!'s Alternative Medicine page, and General Complementary Medicine References. Natural remedies constitute a multimillion dollar industry (Rubin 1997/1998, p. 22). These are sold in pharmacies alongside Western medicines. For example, Longs Drugs had a full page insert in the *San Francisco Chronicle* (1997) advertising dietary supplements such as Ginseng and Ginkgo Biloba and herbal remedies such as Goldenseal.

2. These percentages do not always represent all 40 respondents, because some did not respond to every question.

3. See Kleinman (1996) for a similar discussion on a holistic health center run by baby boomers.

4. Many activists are willing to moderate their discourse to influence mainstream institutions. Hunt, Benford, and Snow (1994, p. 200) explain that when movements interact with these institutions, activists try to "talk their language." For example, a member of Nebraskans for Peace says that their organization realizes that elected officials face financial constraints so activists emphasize the financial "bottom line" when they discuss movement issues with these politicians.

5. Alternative practitioners also learn from these presentations. One alternative practitioner said, "it's not new to me, but there are nuances to everything. When you are a specialist you don't see the whole perspective."

6. The alternative health care movement does not have one identifiable or central organization that unifies the entire movement. Rather, diverse clinics, activist organizations, and individuals compose this diffuse movement. More so than their organizations, alternative health care movement participants are connected through their ideology that they promote through written material and interaction (Gusfield 1994). This ideology allows diverse and often disconnected individuals to have shared meanings, similar experiences, and most importantly, a connection to something larger than their individual participation.

REFERENCES

Aakster, C.W. 1986. "Concepts in Alternative Medicine." *Social Science and Medicine* 22: 265-273.

Alster, K.B. 1989. *The Holistic Health Movement.* Tuscaloosa: The University of Alabama Press.

Baker, B. 1997. "The mind-body connection: Putting the 'faith factor' to work." *AARP Bulletin* 38: 17, 20.

Berliner, H.S., and J.W. Salmon. 1979. "The Holistic Health Movement and Scientific Medicine: The Naked and the Dead." *Socialist Review* 9: 31-52.

Bloom, J. 1987. *HMOs: What They Are, How They Work and Which One is Best for You.* Tucson: The Body Press.

Coulter, H. 1979. "Homeopathic Medicine." Pp. 289-312 in *Ways of Health,* edited by Sobel. New York: Harcourt, Brace and Jovanovich.

Dossey, L. 1996. *Prayer is Good Medicine: How to Reap the Healing Benefits of Prayer.* San Francisco: HarperCollins.

_____. 1989. *Recovering the Soul: A Scientific and Spiritual Search.* New York: Bantam.

_____. 1984. *Beyond Illness: Discovering the Experience of Health.* Boulder, CO. New Science Library.

Easthope, G. 1993. "The Response of Orthodox Medicine to the Challenge of Alternative Medicine in Australia." *The Australian and New Zealand Journal of Sociology* 29: 289-301.

Eisenberg, D.M., R.B. Davis, S.L. Ettner, S. Appel, S. Wilkey, M. Van Rompay, R.C. Kessler. 1998. "Trends in Alternative Medicine Use in the United States, 1990-1997: Results of a Follow-up National Survey." *The Journal of the American Medical Association* 280: 1569-1575.

Eisenberg, D.M., R.C. Kessler, C. Foster, F.E. Norlock, D.R. Calkins, and T.L. Delbanco. 1993. "Unconventional Medicine in the United States: Prevalence, Costs, and Patterns of Use." *New England Journal of Medicine* 328: 246-252.

Gawain, S. with L. King. 1986. *Living in the Light: A Guide to Personal and Planetary Transformation.* San Rafael, CA: Whatever Publishing Inc.

Goldner, M. 1999. "How Alternative Medicine is Changing the Way Consumers and Practitioners look at Quality, Planning of Services and Access in the United States." Pp. 55-74 in *Research in the Sociology of Health Care* (Vol. 16.), edited by J. Kronenfeld. Stamford, CT: JAI Press.

Gray, B. (Ed.). 1986. *For-Profit Enterprise in Health Care.* Washington, D.C.: National Academy Press.

Gusfield, J. 1994. "The Reflexivity of Social Movements: Collective Behavior and Mass Society Theory Revisited." Pp. 58-78 in *New Social Movements: From Ideology to Identity*, edited by E. Larana, H. Johnston, and J. Gusfield. Philadelphia: Temple University Press.

Hunt, S.A., R.D. Benford, and D.A. Snow. 1994. "Identity Fields: Framing Processes and the Social Construction of Movement Identities." Pp. 185-208 in *New Social Movements: From Ideology to Identity*, edited by E. Larana, H. Johnston, and J.R. Gusfield. Philadelphia: Temple University Press.

Jampolsky, G. 1970. *Love is Letting Go of Fear*. Toronto: Bantam.

Jorgensen, D. 1989. *Participant Observation: A Methodology for Human Studies*. Newbury Park: Sage.

Kleinman, S. 1996. *Opposing Ambitions: Gender and Identity in an Alternative Organization*. Chicago: The University of Chicago Press.

Konner, M. 1993. *Medicine at the Crossroads: The Crisis in Health Care*. New York: Pantheon Books.

Kopelman, L., and J. Moskop. 1981. "The Holistic Health Movement: A Survey and Critique." *The Journal of Medicine and Philosophy* 6: 209-235.

Levine, H. 1997. "Natural Medicine: The New Age Rage." *Cosmopolitan*, September, pp. 313-315.

Lowenberg, J. 1989. *Caring and Responsibility: The Crossroads between Holistic Practice and Traditional Medicine*. Philadelphia: University of Pennsylvania Press.

Mattson, P.H. 1982. *Holistic Health in Perspective*. Palo Alto, CA: Mayfield Publishing Company.

Mick, S. 1990. *Innovations in Health Care Delivery: Insights for Organization Theory*. San Francisco: Jossey-Bass.

Northrup, C. 1994. *Women's Bodies, Women's Wisdom: Creating Physical and Emotional Health and Healing*. New York: Bantam Books.

Norwood, R. 1994. *Why Me, Why This, Why Now: A Guide to Answering Life's Toughest Questions*. New York: Carol Southern Books.

Peck, M.S. 1997. *Denial of the Soul: Spiritual and Medical Perspectives on Euthanasia and Mortality*. New York: Random House.

Roof, W.C.. 1993. *A Generation of Seekers: The Spiritual Journeys of the Baby Boom Generation*. San Francisco: Harper.

Rubin, J. 1997/1998. "Letter from the Editor." *Bay Area Naturally*. San Francisco: City Spirit Publications. Fall 1997/Winter 1998, p. 22.

San Francisco Chronicle. 1997. July 23, p. C1.

Schneirov, M., and J.D. Geczik. 1996. "A Diagnosis for our Times: Alternative Health's Submerged Networks and the Transformation of Identities." *The Sociological Quarterly* 37: 627-644.

Scott, R. 1993. "The Organization of Medical Care Services: Toward an Integrated Theoretical Model." *Medical Care Review* 50: 271-302.

Scott, R. and E. Backman. 1990. "Institutional Theory and the Medical Care Sector: Early Theory". Pp. 21-52 in *Innovations in Health Care Delivery: Insights for Organization Theory*, edited by S. Mick. San Francisco: Jossey-Bass.

Starr, P. 1982. *The Social Transformation of American Medicine: The Rise of a Sovereign Profession and the Making of a Vast Industry*. New York: Basic Books, Inc.

Tubesing, D.A. 1979. *Wholistic Health: A Whole-Person Approach to Primary Health Care*. New York: Human Sciences Press.

Van Maanen, J. 1982. "Fieldwork on the Beat: This Being an Account of the Manners and Customs of an Ethnographer in an American Police Department." Pp. 103-149 in *Varieties of Qualitative Research*, edited by J.Van Maanen, J. Dabbs Jr., and R. Faulkner. Beverly Hills: Sage.

Wallis, C. 1996. "Faith and Healing." *TIME*, June 24, pp. 58-64.

Weiss, K. (Ed.). 1984. *Women's Health Care: A Guide To Alternatives*. Reston, VA: Reston Publishing Company, Inc.

Weitz, R. 1996. *The Sociology of Health, Illness, and Health Care: A Critical Approach.* Belmont: Wadsworth Publishing Company.

Williamson, M. 1994. *Illuminata: Thoughts, Prayers and Rites of Passage.* New York: Random House.

Zukav, G. 1989. *The Seat of the Soul.* New York: Simon and Schuster.

EXAMINING THE REAL EFFECT OF PRIOR UTILIZATION ON SUBSEQUENT UTILIZATION

Jin-Yuan Chern, Louis F. Rossiter, and
Thomas T. H. Wan

ABSTRACT

Measures of health status and prior service use have been considered promising predictors of future health expenditures, particularly when used for risk-adjustment models in capitation payment systems. While the use of health status as a future predictor has its difficulty in terms of measurement accuracy and implementation costs, using prior utilization as the base for the calculation of future health expenditures also has its concerns. Based on a three-stage cross-lagged model in a longitudinal study design, this study showed that prior utilization has both a direct and an indirect effect on subsequent utilization. However, the real net effect of prior utilization on subsequent utilization can be overestimated by 25%, if the effect of health status is not taken into account.

Research in the Sociology of Health Care, Volume 17, pages 237-249.
Copyright © 2000 by JAI Press Inc.
All rights of reproduction in any form reserved.
ISBN: 0-7623-0644-0

INTRODUCTION

"From each according to his ability; to each according to his needs," is an ideal give-and-take principle that governed ancient Oriental societies as well as genuinely modern socialism in the nineteenth century. In a health care system with a capitation payment mechanism, the same give-and-take principle holds, through this underlying rationale: People in good health should help those in poor health meet their need for care. Operationally, the funds that individuals pay for care are reallocated to insurers in order to pool the risk within each health plan. Therefore, health plans enrolling disproportionally high-risk persons should receive proportionally high premiums, and vice versa for plans enrolling low-risk persons. The success of funds reallocation from high-risk pools to low-risk pools depends on the adequacy of risk-adjustment modeling that accurately predicts future expenditures for health care. Two of the factors that better predict future expenditures for a group of people are the health status and the prior utilization of health services of the individuals in the group.

Health status has been widely used as a proxy for the need for care in health services research and, recently, in studies predicting future health services utilization and expenditures. In the meantime, prior utilization of health services is shown to have a high predictive power of future expenditures in that prior utilization approaches, especially those with diagnosis- or procedure-based information, have the advantages of a strong clinical foundation and thus have high predictive power (Anderson et al. 1990; Anderson, Steinberg, Holloway, and Cantor 1986; Ash, Porell, Gruenberg, Sawitz, and Beiser 1989; Hayes 1991; Hornbrook and Goodman 1991; Lubitz 1987;Lubitz, Beebe, and Riley 1985;Newhouse, Manning, Keeler, and Sloss 1989; Van de Ven, Van Vliet, Van Barneveld, and Lamers 1994; Van Vliet and Van de Ven 1992; Weiner, Dobson, Maxwell, Coleman, Starfield, and Anderson 1996; Weiner, Starfield, Steinwachs, and Mumford 1991). Some researchers, however, caution that using prior utilization data as the base for the calculation of future health expenditures also has disadvantages: (1) using prior utilization to determine subsequent payments may reward inefficient providers who are unable to control utilization, and penalize those who can control medical costs effectively; (2) when diagnostic or procedural data are used, it is also possible that code-creeping activities may happen, as has been the concern with DRGs (Ash et al. 1989; Ellis et al. 1996; Weiner et al. 1996); an adjustment for prior utilization could also create perverse economic incentives of cost reimbursement (Anderson et al. 1986; McClure 1984; Newhouse 1986); and more importantly, (3) prior utilization data reflect not only an individual's health status, but also the practice styles of previous health plans and providers—a fact that may cause prediction bias to a certain extent (McClure 1984; Robinson, Luft, Gardner, and Morrison 1991). Therefore, the decision of whether or not to include prior utilization in predictive models depends on its real net effect on the prediction of

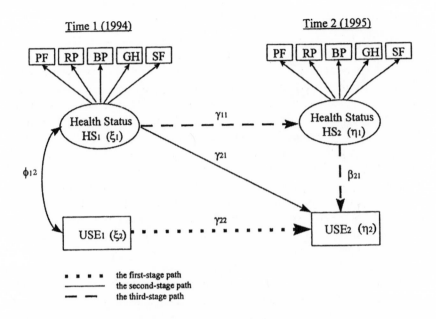

Figure 1. An Analytical Framework for
Services Utilization: A Cross-lagged

subsequent utilization; that is, to what extent does prior utilization have a significant direct effect on subsequent utilization?

In sum, prior utilization may be correlated positively with subsequent service utilization and expenditures, but its predictive power is moderated by the effect on health status that prompted the service use in the first instance. The purpose of this paper is to critically examine the decision of including prior utilization in predictive models for health expenditures based on individual-level data.

In order to accurately examine the essential effects (direct vs. indirect) of prior utilization on subsequent utilization of health services, a study design that appropriately delineates the cause–effect relationships among health status, prior utilization, and subsequent utilization of health services is a crucial necessity. Based on a longitudinal study design, a structural equation modeling technique is used in this study to test a three-stage cross-lagged model of service utilization. Figure 1 depicts the conceptual framework for this study. Prior utilization (USE_1) alone is used first to predict subsequent utilization (USE_2) of health services, prior utilization along with prior health status (HS_1) are then used as future predictors, and finally, subsequent health status (HS$_2$) is also included in predictive models. Figure 1 is elaborated in more detail below.

METHODS

Data Sources and Sample Selection

Data were collected through two waves of health status mail survey (at the end of each year, 1994 and 1995, respectively). All policyholders at Virginia's largest insurer were eligible for being sampled if they met the following criteria:

1. Employed at either one public-sector agency with a Point of Service plan (i.e., POS) or at one private-sector company with Preferred Provider Organization (PPO) or Traditional Fee for Service (TFFS), since January 1993
2. Had continuous group coverage at Trigon BC/BS since January 1993
3. Not a member of any other Trigon BC/BS HMO product
4. Being between ages 18 and 64 in July 1994

Policyholders were randomly selected from the membership list of the insurer. The total number of policyholders to be surveyed in the first wave (year 1994) was 14,331. Among the 8,574 policyholders (59.84%) who responded to the first wave survey, 5,640 (65.78%) responded to the second wave survey (year 1995). In recognition of the restrictive data requirements for a longitudinal study, observations with missing data on any variables (1,385 total, or 24.6%) were eliminated. Thus, this study is based on a panel of 4,255 policyholders.

Measure of Health Status

Among different approaches to measuring health status, the use of MOS 36-Item Short-Form Health Survey, or abbreviated as SF-36 (Ware and Sherbourne 1992), has gained popularity for both research and administrative purposes (Hornbrook and Goodman 1995, 1996; Pai and Wan 1997). Briefly described, the SF-36 Health Survey measures health status in eight dimensions, each measuring one health concept: physical functioning (PF), role limitations due to physical health problems (RP), bodily pain (BP), general health (GH), vitality (VT), social functioning (SF), role limitations due to emotional problems (RE), and mental health (i.e., psychological distress and psychological well-being) (MH). A more detailed description of the items in each scale is provided in the SF-36 Health Survey manual (Ware, Snow, Kosinske, and Gandek 1993). Theoretically, the first three dimensions (i.e. PF, RP, and BP) measure mainly physical health status; the last two (i.e., RE and MH) measure mental health status; and, the other three dimensions (i.e., BP, GH, and VT) somewhat cross the two principal components (Ware, Kosinski, and Keller 1995; Pai and Wan 1997). The validity and reliability of SF-36 have been partially evaluated and shown to be acceptable as measures of general health and the health-related quality of life (McHorney, Ware, and Faczek 1993; McHorney, Ware, Rachel, and Sherbourne 1994; Ware and Sherbourne 1992; Ware et al. 1993), although those tests were mainly conducted within cross-sectional study windows. In one study comparing the predictive performance of risk-adjustment methods, only PF, RP, BP, and GH scales from the SF-36 Health Survey instrument were retained for analysis, as determined by the predictive accuracy of indicators (Fowles, Weiner, Knutson, Fowler, Tucker, and Ireland 1996). Still, in previous studies (Chern 1997), a parsimonious measurement model of health status as indicated by five scales (including PF, RP, BP, GH, and SF) was shown to perform as well as the comprehensive eight-scale model in conceptualizing multidimensional health status. More importantly, the parsimonious measurement model was further proven to be highly stable over time, while the original eight-scale model failed to demonstrate its stability. Therefore, the five-scale model of health status will be used for this study.

Health Services Utilization

The total annual charges for health care expenditures, extracted from the claims and membership information system at the insurer, were used to represent the amount of services used over one year by each policyholder. Expenditures for out-of-plan utilization are included in the totals, to the extent that the plans reimbursed the out-of-plan utilization.

In this study, the total annual charges included services rendered at physician offices and in hospitals or other facilities. In the meantime, the total charges are used for analysis because they reflect the market prices of health services, as opposed to the "paid amounts," which indicate transaction prices.

Three-Stage Cross-lagged Model

The statistical analysis is based mainly on structural equation modeling (SEM) technique (Bollen 1989; Long 1983a, 1983b; Jöreskog and Sörbom 1989; Hayduk 1987), with the aid of the newly developed software package AMOS (Analysis of MOment Structure) (Arbuckle 1997). Briefly, a structural equation model evaluates whether the proposed causal relationship is consistent with the actual patterns found among variables in the empirical data. The technique is considered appropriate in situations where multiple constructs and indicators (e.g., health status) are taken into account at the same time.

Although prior utilization has been shown to be highly correlated with subsequent utilization, it is also presumed that use of health services should improve health status and thus reduce subsequent services utilization. In order to test for the essential effects (direct vs. indirect) of prior utilization on subsequent utilization of health services when taking health status into account, a three-stage approach was adopted: {bl}

1. Utilization data from years 1994 (Time 1) and 1995 (Time 2) were used.
2. Utilization data from years 1994 (Time 1) and 1995 (Time 2) along with the first-wave health status from year-end 1994 (Time 1) were included in one model simultaneously.
3. The second-wave health status from year-end 1995 (Time 2) was added to the preceding model and became a complete model.

The structural equation model for this study was diagrammatically illustrated in Figure 1.

At the first stage, the direct effect of prior utilization on subsequent utilization is indicated by the regression coefficient γ_{22}. When measures of health status were also included (the second- and the third-stage), the real direct effect of prior utilization on subsequent utilization was re-examined by comparing the regression coefficients between the two stages.

Moreover, the indirect effect of prior utilization on subsequent utilization, through health status, is indicated in the second stage by the product of two coefficients: θ_{12} (between USE_1 [ξ_2] and $HS_1[\xi_1]$), γ_{21} (between HS_1 and USE_2 [η_2]). It is indicated in the third stage by the product of three coefficients: θ_{12} (between USE_1 and HS_1), γ_{11} (between HS_1 and HS_2 [η_1]), and β_{21} (between HS_2 and USE_2).

In summary, based on a panel model incorporating two waves of expenditures data and two measures of health status, the real effects (direct vs. indirect) of prior utilization on subsequent utilization are hypothesized as follows. {bl}

Hypothesis A (Ha):   Prior services utilization has a positive direct effect on subsequent utilization.

Hypothesis B (Hb):   Prior services utilization (1994) has a positive indirect effect on subsequent utilization (1995), through health status 1994.

Hypothesis C (Hc):   Prior services utilization (1994) has a positive indirect effect on subsequent utilization (1995), through health status 1994 and 1995.

The direct effect is examined by comparing the regression coefficient between prior utilization (1994) and subsequent utilization (1995), when health status is not included, to the corresponding parameter coefficient, when health status is included. The indirect effect (through health status) is computed by multiplying parameter coefficients between prior utilization (i.e., 1994) and subsequent utilization (i.e., 1995) through two sets of path, as demonstrated in Figure 1.

RESULTS

Descriptive Statistics of Study Sample

As of July 1994, the mean age of the study sample was slightly above 45. About 63% were males. The majority of the subjects were White/Caucasian (85%). For most, their highest education was between a high school and a college degree, with a few having less than a high school diploma (6%), and some with postgraduate education (16%). Three-quarters of the study sample were married. More than two-thirds (67.4%) were enrolled in a Point of Service (POS) health plan, 24.3% were in a Preferred Provider Organization (PPO), and 8.3% were in traditional fee-for-service (TFFS). The household income fell mainly into categories between $20,000 to $59,999 (62.4%), with 30.4% at more than $60,000 and 7% at less than $20,000. More than half (57%) of the study sample were white-collar workers (executives/managers, professionals, and administrators/clerical staff).

Table 1 lists health status as measured by five indicators from the two-wave survey. Among these health concepts, four scales (PF, RP, BP, and SF) define health as the absence of limitation or disability. For these scales, a highest score of 100 represents no limitations or disabilities observed. For the other scale (GH), a score in the mid-range indicates no limitations or disabilities. A score of 100, in contrast, is achieved only when respondents report positive states and evaluate their health favorably (Ware et al. 1993). Overall, the health status scores of the study group were above the norms for the general U.S. population (see Table 10.1 in Ware et al. 1993). All five dimensions of health status showed a positively skewed distribution, indicating a relatively healthy study sample. Except for physical functioning (PF) and general health perceptions (GH), all other dimensions showed an improved change between the first- and second-wave survey, and all changes were within one point of the scores.

Table 2 shows the total annual health services expenditures of the study sample in years 1994 and 1995. As expected, health services expenditures of the panel

Table 1.   Health Status Scores on SF-36: Five Dimensions

SF-36 Health Concepts	First Wave (1994) (N = 4,255)		Second Wave (1995) (N = 4,255)		U.S. Norms (N = 2,474)	
	Mean	*S.D.*	*Mean*	*S.D.*	*Mean*	*S.D.*
Physical functioning (PF)	87.43	18.78	87.39	19.33	84.15	23.28
Role-physical (RP)	85.50	29.52	85.83	29.74	80.96	34.00
Bodily pain (BP)	75.28	22.40	75.86	22.08	75.15	23.69
General health (GH)	73.36	19.26	73.12	19.31	71.92	20.34
Social functioning (SF)	87.91	19.31	88.07	19.58	83.28	22.69

Note: *based on general U.S. population (Ware et al., 1993, Table 10.1).

Table 2. Health Services Expenditures, 1994—1995
(*N* = 4,255)

	Year 1994		Year 1995	
Services Use	*Mean*	*S.D. or %*	*Mean*	*S.D. or %*
Total charges	$2,787	$8,002	$2,656	$9,706
No. of users	3,796	89.2%	3,754	88.2%
No. of nonuser	459	10.8%	501	11.8%

Table 3. Results of Three-Stage Cross-Lagged Model—Real Effect
of Prior Utilization

Coefficients	*The First-Stage*	*The Second-Stage*	*The Third-Stage*
θ_{11}	N.A.	-0.29^{***}	-0.29^{***}
γ_{11}	N.A.	N.A.	0.83^{***}
γ_{21}	N.A.	-0.17^{***}	0.04
γ_{22}	0.20^{***}	0.15^{***}	0.15^{***}
β_{21}	N.A.	N.A.	-0.25^{***}
Indirect Effect	N.A.	0.05	0.06
SMC (Subsequent Expenditues)	0.04	0.07	0.09
Overall Model Fit			
Goodness-of-fit index	N.A.	1.00	0.99
Adjusted goodness-of-fit index	N.A.	0.99	0.99
Root mean square error of approx.	N.A.	0.03	0.03
P_close value	N.A.	1.00	1.00
Critical N	N.A.	2,065	1,165

Notes: ***0.001 level of significance
N.A. = not applicable
correlated errors not shown

sample steadily declined between the years; on average, health services expenditures per capita were $2,787 and $2,656 in calendar years 1994 and 1995, respectively. About 89.2% of the study sample used a certain amount of health services in 1994; about 88.2% did so in 1995. That is, in each year only about 10% to 12% of the study sample did not use any health services.

Direct vs. Indirect Effects

Table 3 shows the maximum likelihood estimates of parameters for the structural equation models.

All direct effects were statistically significant at the 0.001 level: 0.20 at the first stage (excluding health status), 0.15 at the second stage (including the first-wave health status), and 0.15 at the third stage (including both the first- and the second-wave health status). In other words, the hypothesis Ha is supported. Moreover, the findings show that the direct effect of prior utilization on subsequent utilization decreases by 25% (from 0.20 to 0.15), when health status is taken into account (at both the second- and the third-stage).

When the first-wave health status was included (the second stage), the indirect-joint effect of prior utilization on subsequent utilization is the cross-product of −0.29 by −0.17, or 0.05. When both waves of health status were included (the third stage), the indirect-joint effect is the cross-product of −0.29 by 0.83 by −0.25, or 0.06. In other words, prior utilization has a positive indirect effect on subsequent utilization when only the first wave or both waves of health status are taken into consideration. Therefore, the hypotheses Hb and Hc are both supported.

Moreover, at the first stage, the direct effect of prior health status on subsequent utilization was statistically significant at the 0.001 level with a γ_{21} value of −0.17. When both waves of health status were included (the second stage), the cross-lagged indirect effect of prior health status on subsequent utilization became insignificant with γ_{21} being 0.04; however, the concurrent direct effect of health status on utilization was statistically significant at the 0.001 level with γ_{21} equaling −0.25. In the meantime, the direct effect of prior health status on subsequent health status was statistically significant at the 0.001 level with γ_{11} equaling 0.83.

Also included in Table 3 are coefficient of determination and five overall model fit indices. Briefly, the coefficient of determination (COD, or parallel to R^2 in general regression models, with values between 0 and unity) measures the strength of several relationships jointly (Jöreskog and Sörbom 1989). Goodness-of-fit (GFI, ranging from 0 to 1) is a measure of the amount of variances and covariances jointly accounted for by the model; the larger, the better. Adjusted goodness-of-fit (AGFI) is a measure of goodness-of-fit while taking into account the degrees of freedom available. Root mean squared error of approximation (RMSEA) measures the degree of model adequacy based on population discrepancy in relation to degrees of freedom; a value less than 0.05 (or 0.08) is preferred (acceptable). P_close is a "p-value" for testing the null hypothesis that RMSEA is equal to or less than 0.05. A P_close value equal to or greater than 0.05 indicates a close model fit. Hoelter's critical N (CN) indicates the largest sample size for which one would accept the hypothesis that a model is correct. Usually, a CN equal to or greater than 200 is needed.

The coefficients of determination for corresponding model of health expenditures are as follows: 0.04 at the first stage, when health status is excluded; 0.07 at the second stage, when the first-wave health status is used; and 0.09 at the third stage, when both waves of health status are included. In other words, a predictive model based on health status increases the percentage of variance captured in the model.

The overall evaluation of model fit is satisfactory in view of the following results: in the second-stage model, GFI=1.00, AGFI=0.99, RMSEA=0.03, P_close=1.00, and CN=2,065; in the third-stage model, GFI=0.99, AGFI=0.99, RMSEA=0.03, P_close=1.00, and CN=1,166. All correlations of measurement errors are below 0.5 and statistically significant at the 0.001 level. There is no relevant information for the first-stage model since it is a saturated (just-identified) model.

CONCLUSIONS

The direct effect of prior utilization on subsequent utilization is positive in each of the three stages, as is traditionally expected. In other words, the positive coefficients between prior utilization and subsequent utilization (γ_{22}=0.20, p<0.001, in the first stage, and γ_{22}=0.15, p<0.001, in both the second- and the third-stage model) indicate a "traditional effect" of prior utilization on subsequent utilization: Higher (lower) prior utilization has a higher (lower) utilization in the subsequent year. The indirect effects in both the second- and the third-stage models are positive.

The negative correlation coefficient between prior year health status and prior year health expenditures (θ_{12}=−0.29, p<0.001) suggests an effect of "concurrent reflection" between health status and health services utilization; that is, higher (lower) health expenditures in one year "reflect" a poorer (better) health status in the same year.

The negative regression coefficient between prior health status and subsequent health expenditures (i.e., γ_{21}=−0.17, p<0.001) in the second-stage model suggests an expected effect of "predicting" by health status on services utilization; that is, poorer (better) health status "predicts" potentially higher (lower) health services expenditures. In the meantime, the change in regression weights between prior health status and subsequent health expenditures (i.e., γ_{21}=−0.17 in the second-stage model, and γ_{21}=0.04 in the third-stage model) suggests a recent effect of health status on health expenditures, as indicated by the significant negative coefficient between subsequent health status and subsequent health expenditures (β_{21}=−0.25, p<0.001). On the other hand, the positive regression coefficient between two measures of health status (γ_{11}=0.83, p<0.001, in the third-stage model) suggests a high stability of health in the working population enrolled in the health plans.

As argued by many studies, prior utilization, especially that caused by chronic disease, may continue for a long period of time and may consume more services in the subsequent years (Anderson et al. 1986; Ash et al. 1989; Howland, Stokes, Crane, and Belanger 1987; Newhouse et al. 1989; Schauffler, Howland, and Cobb 1992; Van Vliet and Van de Ven 1993). In other words, health expenditures are somewhat a "tradition" of prior utilization. The findings reflect the undesirable phenomenon in the real world: The more health services one uses, the more health services one will use in the future. Under the pressure of managed care, utilization management needs to pay attention to enrollees' utilization patterns so that proper disease management or case management of high users of health services can be implemented.

On the other hand, the real direct effect of prior utilization on subsequent utilization decreases by 25% when health status is taken into account. In other words, without controlling for the joint effect of health status, the direct effect of prior utilization on subsequent utilization is overestimated by 25%. Therefore, in capitation payment systems where prior utilization is used as a determinant of future health expenditures in the risk-adjustment models, a close examination of regression weights for prior utilization is needed. Furthermore, it is imperative to examine the joint effects of prior health and health services use on future use of health services. To put it differently, in a health care market where capitated payment is used, either by insurers to pay health services providers or by third-party payers buying insurance, slight changes in the capitation formula predictors could create greatly excessive profits (or losses) for insurers and providers and would significantly affect access to care for those insured as well. This consideration is critical for building a fare payment system if capitation is to be fully implemented. For example, in a health market with unfair rate settings, insurers might opt for "cream skimming" strategies to keep their costs down or to achieve excessive profits; insureds, on the other hand, might respond with "adverse selection" so as not to be excluded from insurance markets. Either way, it will result in the collapse of a capitation payment system due to the inappropriate design of risk-adjustment modeling.

Furthermore, rising health care costs in the United States may be attributed to several factors: aging of the population; the spread of health insurance; rising income; new technologies; diminished tolerance of uncertainty among patients, physicians, and third-party payers; and so on (Kravitz and Greenfield 1995; Newhouse 1993). From the perspective of capitated payment systems, however, the fundamental component of escalating costs is still the utilization of health services. Therefore, it is essential to understand the determinants of service utilization so that health care costs can be contained effectively. As demonstrated in this study, to accurately reflect service utilization and bring health costs under control, the importance of concomitant effect of prior utilization and health status on future health and health services use cannot be overemphasized, especially when we try to understand the mechanisms that may affect capitated payment systems.

There is clearly a pressing need for understanding the interplay between individual and societal factors that affect the variation in health services use of HMO enrollees. The use of a panel study design enables us to gain better knowledge of the concomitant effect of prior health status and health services use on future health. More importantly, a well-designed risk-adjustment model with appropriate concern with prior health services use and health status is inescapable to warrant the successful implementation of capitation payment systems.

ACKNOWLEDGMENTS

This study is part of a larger project General Health Study that was funded by Trigon BlueCross/BlueShield of Virginia and undertaken by Williamson Institute for Health Studies at Virginia Commonwealth University, Medical College of Virginia campus.

REFERENCES

Anderson, G.F., E.P. Steinberg, J. Holloway, and J.C. Cantor, 1986. "Paying for Care: Issues and Options in Setting Capitation Rates." *The Milbank Quarterly* 64(4): 548-565.

Anderson, G.F., E.P. Steinberg, N.R. Powe, S. Antebi, J. Whittle, S. Horn, and R. Herbert. 1990. "Setting Payment Rates for Capitated Systems: A Comparison of Various Alternatives." *Inquiry* 27: 225-233.

Arbuckle, J.L. (1997). *AMOS Users' Guide.* Chicago, IL: SPSS Inc.

Ash, A., F. Porell, L. Gruenberg, E. Sawitz, and A. Beiser. 1989. "Adjusting Medicare Capitation Payments Using Prior Hospitalization Data." *Health Care Financing Review* 10(4): 17-29.

Bollen, K. 1989. *Structural Equations with Latent Variables.* New York: John Wiley and Sons, Inc.

Chern, J.Y. (1997). *Determinants of Health Services Expenditures: A Longitudinal Study.* Dissertation, Virginia Commonwealth University, Richmond, VA.

Ellis, R.P., G.C. Pope, L. I. Iezzoni, J.Z. Ayanian, D.W. Bates, H. Burstin, and A.S. Ash. 1996. "Diagnosis-Based Risk Adjustment for Medicare Capitation Payments." *Health Care Financing Review* 17(3): 101-129.

Fowles, J.B., J.P. Weiner, D. Knutson, E. Fowler, A.M. Tucker, and M. Ireland. 1996. "Taking Health Status into Account When Setting Capitation Rates: A Comparison of Risk-Adjustment Methods." *Journal of the American Medical Association* 276(16): 1316-1321.

Jöreskog, K., and D. Sörbom. 1989. *LISREL 7: A Guide to the Program and Applications.* (2nd ed.). Chicago, IL. SPSS Inc.

Hayduk, L.A. 1987. *Structural Equation Modeling: Essentials and Advances.* Baltimore, MD: The Johns Hopkins University Press.

Hayes, S.T. 1991. "Demographic Risk Factors Derived from HMO Data." Pp. 177-196 in *Advances in Health Economics and Health Services Research*, edited by R.M. Scheffler and L.F. Rossiter. London: JAI Press.

Hornbrook, M.C., and M.J Goodman. 1991. "Health Plan Case Mix: Definition, Measurement, and Use." Pp. 111-148 in *Advances in Health Economics and Health Services Research*, edited by R.M. Scheffler and L.F. Rossiter. London: JAI Press.

_____. 1995. "Assessing Relative Health Plan Risk with the RAND-36 Health Survey." *Inquiry* 32: 56-74.

_____. 1996. "Chronic Disease, Functional Health Status, and Demographics: A Multi-Dimensional Approach to Risk Adjustment." *Health Services Research* 31(3): 283-307.

Howland, J., J.I. Stokes, S.C. Crane, and A.J Belanger. 1987. "Adjusting Capitation Using Chronic Disease Risk Factors: A Preliminary Study." *Health Care Financing Review* 9(2): 15-23.

Kravitz, R.L., and S. Greenfield. 1995. "Variations in Resource Utilization among Medical Specialties and Systems of Care." Pp. 431-455 in *Annual Review of Public Health*, edited by G.S. Omenn, J.E. Fielding, and L.B. Lave. Palo Alto, CA: Annual Reviews.

Long, J.S. 1983a. *Confirmatory Factor Analysis: A Preface to LISREL.* Newbury Park, CA: Sage Publications, Inc.

_____. 1983b. *Covariance Structure Models: An Introduction to LISREL.* Newbury Park, CA: Sage Publications, Inc.

Lubitz, J. 1987. "Health Status Adjustments for Medicare Capitation." *Inquiry* 24(Winter): 362-375.

Lubitz, J., J. Beebe, and G. Riley. 1985. "Improving the Medicare HMO Payment Formula to Deal with Biased Selection." Pp. 101-122 in *Advances in Health Economics and Health Services Research*, edited by R.M. Scheffler and L.F. Rossite. London: JAI Press.

McClure, W. 1984. "On the Research Status of Risk-Adjusted Capitation Rates." *Inquiry* 21: 205-213.

McHorney, C., J.E. Ware, Jr., and A. Faczek. 1993. "The MOS 36-Item Short-Form Health Survey (SF-36): II. Psychometric and Clinical Tests of Validity in Measuring Physical and Mental Health Constructs." *Medical Care* 31(3): 247-263.

McHorney, C., J.E. Ware, Jr., J. Rachel, and C. Sherbourne. 1994. "The MOS 36-Item Short-Form Health Survey (SF-36): III. Tests of Data Quality, Scaling Assumptions, and Reliability Across Diverse Patient Groups." *Medical Care* 32(1): 40-66.

Newhouse, J.P. 1986. "Rate Adjusters for Medicare under Capitation." *Health Care Financing Review, Annual Supplement* 45-55.

_____. 1993. "An Iconoclastic View of Health Costs Containment." *Health Affairs* 12(Suppl.): 152-171.

Newhouse, J.P., W.G. Manning, E.B. Keeler, and E.M Sloss. 1989. "Adjusting Capitation Rates Using Objective Health Measures and Prior Utilization." *Health Care Financing Review* 10(3): 41-54.

Pai, C.W. and T.T.H. Wan. 1997. "Confirmatory Analysis of Health Outcome Indicators: The 36-Item Short-Form Health Survey (SF-36)." *Journal of Rehabilitation Outcomes Research* 1(2):48-59.

Robinson, J.C., H.S. Luft, L.B. Gardner, and E.M Morrison. 1991. "A Method for Risk-Adjusting Employer Contributions to Competing Health Insurance Plans." *Inquiry* 28(2): 107-116.

Schauffler, H.H., J. Howland, and J. Cobb. 1992. "Using Chronic Disease Risk Factors to Adjust Medicare Capitation Payments." *Health Care Financing Review* 14(1): 79-90.

Van de Ven, W., R. Van Vliet, E. Van Barneveld, and L.M. Lamers. 1994. "Risk-Adjusted Capitation: Recent Experiences in the Netherlands." *Health Affairs* 13(Winter): 120-136.

Van Vliet, R., and W. Van de Ven. 1992. "Towards a Capitation Formula for Competing Health Insurers: An Empirical Analysis." *Social Science and Medicine* 34(9): 1035-1048.

_____. 1993. "Capitation Payments Based on Prior Hospitalizations." *Journal of Health Economics* 2, 177-188.

Ware, J.E., and C.D. Sherbourne. 1992. "The MOS 36-Item Short-Form Health Survey (SF-36): I. Conceptual Framework and Item Selection." *Medical Care* 30(6): 473-481.

Ware, J.E., M. Kosinski, and S.D. Keller. 1995. *SF-12: How to Score the SF-12 Physical and Mental Summary Scales* (2nd ed.). Boston, MA: The Health Institute, New England Medical Center.

Ware, J.E., K.K. Snow, M. Kosinske, and B. Gandek. 1993. *SF-36 Health Survey: Manual and Interpretation Guide*. Boston, MA: Nimrod Press.

Weiner, J.P., A. Dobson, S.L. Maxwell, K. Coleman, B.H. Starfield, and G. Anderson. 1996. "Risk-Adjusted Capitation Rates Using Ambulatory and Inpatient Diagnoses." *Health Care Financing Review* 17(3): 77-99.

Weiner, J.P., B.H. Starfield, D.M. Steinwachs, and L.M. Mumford. 1991. "Development and Application of a Population-Oriented Measure of Ambulatory Care Case-Mix." *Medical Care* 29(5): 452-472.